GED Subject Test

Mathematics

Student Practice Workbook

+ Two Realistic GED Math Tests

Math Notion

www.MathNotion.com

GED Subject Test – Mathematics

GED Subject Test – Mathematics

GED Subject Test Mathematics

Published in the United State of America By

The Math Notion

Web: WWW.MathNotion.com

Email: info@Mathnotion.com

Copyright © 2021 by the Math Notion. All rights reserved. No part of this publication may be reproduced, stored in a retrieval system, or transmitted in any form or by any means, electronic, mechanical, photocopying, recording, scanning, or otherwise, except as permitted under Section 107 or 108 of the 1976 United States Copyright Ac, without permission of the author.

All inquiries should be addressed to the Math Notion.

ISBN: 978-1-63620-038-5

GED Subject Test – Mathematics

The Math Notion

Michael Smith has been a math instructor for over a decade now. He launched the Math Notion. Since 2006, we have devoted our time to both teaching and developing exceptional math learning materials. As a test prep company, we have worked with thousands of students. We have used the feedback of our students to develop a unique study program that can be used by students to drastically improve their math scores fast and effectively. We have more than a thousand Math learning books including:

– HiSET Math Prep
– TABE Math Prep
– TASC Math Prep
– Accuplacer Math Prep
– Common Core Math Prep
–many Math Education Workbooks, Study Guides, Practice and Exercise Books

As an experienced Math test preparation company, we have helped many students raise their standardized test scores—and attend the colleges of their dreams: We tutor online and in person, we teach students in large groups, and we provide training materials and textbooks through our website and through Amazon.

You can contact us via email at:

info@Mathnotion.com

GED Subject Test – Mathematics

Get the Targeted Practice You Need to Ace the GED Math Test!

GED Subject Test - Mathematics includes easy-to-follow instructions, helpful examples, and plenty of math practice problems to assist students to master each concept, brush up their problem-solving skills, and create confidence.

The GED math practice book provides numerous opportunities to evaluate basic skills along with abundant remediation and intervention activities. It is a skill that permits you to quickly master intricate information and produce better leads in less time.

Students can boost their test-taking skills by taking the book's two practice GED Math exams. All test questions answered and explained in detail.

Important Features of the GED Math Book:

- A **complete review** of GED math test topics,
- Over 2,500 practice problems covering all topics tested,
- The most important concepts you need to know,
- Clear and concise, easy-to-follow sections,
- Well designed for enhanced learning and interest,
- Hands-on experience with all question types
- **2 full-length practice tests** with detailed answer explanations
- Cost-Effective Pricing

Powerful math exercises to help you avoid traps and pacing yourself to beat the GED test. Students will gain valuable experience and raise their confidence by taking math practice tests, learning about test structure, and gaining a deeper understanding of what is tested on the GED Math. If ever there was a book to respond to the pressure to increase students' test scores, this is it.

GED Subject Test – Mathematics

WWW.MathNotion.COM

… So Much More Online!

✓ FREE Math Lessons

✓ More Math Learning Books!

✓ Mathematics Worksheets

✓ Online Math Tutors

For a PDF Version of This Book

Please Visit WWW.MathNotion.com

GED Subject Test – Mathematics

Contents

Chapter 1 : Integers and Number Theory 11

    Rounding 12

    Whole Number Addition and Subtraction 13

    Whole Number Multiplication and Division 14

    Rounding and Estimates 15

    Adding and Subtracting Integers 16

    Multiplying and Dividing Integers 17

    Order of Operations 18

    Ordering Integers and Numbers 19

    Integers and Absolute Value 20

    Factoring Numbers 21

    Greatest Common Factor 22

    Least Common Multiple 23

    Answers of Worksheets 24

Chapter 2 : Fractions and Decimals 27

    Simplifying Fractions 28

    Adding and Subtracting Fractions 29

    Multiplying and Dividing Fractions 30

    Adding and Subtracting Mixed Numbers 31

    Multiplying and Dividing Mixed Numbers 32

    Adding and Subtracting Decimals 33

    Multiplying and Dividing Decimals 34

    Comparing Decimals 35

    Rounding Decimals 36

    Answers of Worksheets 37

Chapter 3 : Proportions, Ratios, and Percent 40

    Simplifying Ratios 41

    Proportional Ratios 42

    Similarity and Ratios 43

    Ratio and Rates Word Problems 44

WWW.MathNotion.Com

GED Subject Test – Mathematics

Percentage Calculations 45
Percent Problems 46
Discount, Tax and Tip 47
Percent of Change 48
Simple Interest 49
Answers of Worksheets 50

Chapter 4 : Exponents and Radicals Expressions 53
Multiplication Property of Exponents 54
Zero and Negative Exponents 55
Division Property of Exponents 56
Powers of Products and Quotients 57
Negative Exponents and Negative Bases 58
Scientific Notation 59
Square Roots 60
Simplifying Radical Expressions 61
Answers of Worksheets 62

Chapter 5 : Algebraic Expressions 65
Simplifying Variable Expressions 66
Simplifying Polynomial Expressions 67
Translate Phrases into an Algebraic Statement 68
The Distributive Property 69
Evaluating One Variable Expressions 70
Evaluating Two Variables Expressions 71
Combining like Terms 72
Answers of Worksheets 73

Chapter 6 : Equations and Inequalities 75
One–Step Equations 76
Multi–Step Equations 77
Graphing Single–Variable Inequalities 78
One–Step Inequalities 79
Multi-Step Inequalities 80
Systems of Equations 81
Systems of Equations Word Problems 82

GED Subject Test – Mathematics

Answers of Worksheets 83

Chapter 7 : Linear Functions 87
Finding Slope 88
Graphing Lines Using Line Equation 89
Writing Linear Equations 90
Graphing Linear Inequalities 91
Finding Midpoint 92
Finding Distance of Two Points 93
Answers of Worksheets 94

Chapter 8 : Polynomials 97
Writing Polynomials in Standard Form 98
Simplifying Polynomials 99
Adding and Subtracting Polynomials 100
Multiplying Monomials 101
Multiplying and Dividing Monomials 102
Multiplying a Polynomial and a Monomial 103
Multiplying Binomials 104
Factoring Trinomials 105
Operations with Polynomials 106
Answers of Worksheets 107

Chapter 9 : Functions Operations and Quadratic 111
Evaluating Function 112
Adding and Subtracting Functions 113
Multiplying and Dividing Functions 114
Composition of Functions 115
Quadratic Equation 116
Solving Quadratic Equations 117
Quadratic Formula and the Discriminant 118
Graphing Quadratic Functions 119
Answers of Worksheets 120

Chapter 10 : Geometry and Solid Figures 123
Angles 124
Pythagorean Relationship 125

WWW.MathNotion.Com

GED Subject Test – Mathematics

    Triangles 126

    Polygons 127

    Trapezoids 128

    Circles 129

    Cubes 130

    Rectangular Prism 131

    Cylinder 132

    Pyramids and Cone 133

    Answers of Worksheets 134

Chapter 11 : Statistics and Probability 137

    Mean and Median 138

    Mode and Range 139

    Times Series 140

    Stem–and–Leaf Plot 141

    Pie Graph 142

    Probability Problems 143

    Answers of Worksheets 144

Chapter 12 : GED Test Review 147

    GED Practice Test 1 151

    GED Practice Test 2 167

Chapter 13 : Answers and Explanations 183

    Answer Key 183

    Practice Test 1 185

    Practice Test 2 193

GED Subject Test – Mathematics

Chapter 1:
Integers and Number Theory

Topics that you'll practice in this chapter:

- ✓ Rounding
- ✓ Whole Number Addition and Subtraction
- ✓ Whole Number Multiplication and Division
- ✓ Rounding and Estimates
- ✓ Adding and Subtracting Integers
- ✓ Multiplying and Dividing Integers
- ✓ Order of Operations
- ✓ Ordering Integers and Numbers
- ✓ Integers and Absolute Value
- ✓ Factoring Numbers
- ✓ Greatest Common Factor (GCF)
- ✓ Least Common Multiple (LCM)

"Wherever there is number, there is beauty." –Proclus

GED Subject Test – Mathematics

Rounding

✎ **Round each number to the nearest ten.**

1) 42 = ___   5) 19 = ___   9) 48 = ___

2) 88 = ___   6) 25 = ___   10) 81 = ___

3) 24 = ___   7) 93 = ___   11) 58 = ___

4) 57 = ___   8) 71 = ___   12) 87 = ___

✎ **Round each number to the nearest hundred.**

13) 198 = ___   17) 321 = ___   21) 580 = ___

14) 387 = ___   18) 433 = ___   22) 868 = ___

15) 816 = ___   19) 579 = ___   23) 480 = ___

16) 101 = ___   20) 825 = ___   24) 287 = ___

✎ **Round each number to the nearest thousand.**

25) 1,382 = ___   29) 9,099 = ___   33) 52,866 = ___

26) 3,420 = ___   30) 22,980 = ___   34) 85,190 = ___

27) 4,254 = ___   31) 45,188 = ___   35) 70,990 = ___

28) 6,861 = ___   32) 16,808 = ___   36) 26,869 = ___

WWW.MathNotion.Com

GED Subject Test – Mathematics

Whole Number Addition and Subtraction

✎ Find the sum or subtract.

1) 1,240 + 658 = _____

2) 3,458 − 544 = _____

3) 2,259 − 752 = _____

4) 1,990 + 324 = _____

5) 3,088 + 229 = _____

6) 2,354 + 1,009 = _____

7) 2,855 + 4,455 = _____

8) 5,112 + 4,004 = _____

9) 4,822 − 2,007 = _____

10) 8,380 − 5,288 = _____

11) 3,227 + 4,150 = _____

12) 7,702 − 4,331 = _____

✎ Find the missing number.

13) 720 + ____ = 1,360

14) 2,115 − ____ = 1,103

15) ____ + 3,105 = 6,200

16) 5,250 − 3,280 = ____

17) 8,020 + ____ = 8,990

18) 7,302 − 4,700 = ____

WWW.MathNotion.Com

GED Subject Test – Mathematics

Whole Number Multiplication and Division

✎ **Calculate each product.**

1) 42 × 13 = _____

2) 70 × 15 = _____

3) 40 × 14 = _____

4) 22 × 20 = _____

5) 110 × 11 = _____

6) 150 × 16 = _____

✎ **Find the missing quotient.**

7) 564 ÷ 6 = _____

8) 270 ÷ 3 = _____

9) 640 ÷ 8 = _____

10) 450 ÷ 9 = _____

11) 112 ÷ 7 = _____

12) 1,260 ÷ 9 = _____

13) 3,000 ÷ 10 = _____

14) 2,400 ÷ 8 = _____

15) 3,200 ÷ 40 = _____

16) 6,300 ÷ 90 = _____

✎ **Calculate each problem.**

17) 400 ÷ 5 = N, N = ___

18) 2,500 ÷ 10 = N, N = ___

19) N ÷ 3 = 150 , N = ___

20) 42 × N = 252, N = ___

21) 660 ÷ N = 330, N = ___

22) N × 8 = 336, N = ___

GED Subject Test – Mathematics

Rounding and Estimates

✏️ **Estimate the sum by rounding each number to the nearest ten.**

1) $13 + 22 =$ _____

2) $71 + 23 =$ _____

3) $61 + 58 =$ _____

4) $56 + 85 =$ _____

5) $368 + 249 =$ _____

6) $330 + 903 =$ _____

7) $471 + 293 =$ _____

8) $1{,}950 + 2{,}655 =$ _____

✏️ **Estimate the product by rounding each number to the nearest ten.**

9) $32 \times 71 =$ _____

10) $12 \times 33 =$ _____

11) $31 \times 83 =$ _____

12) $19 \times 11 =$ _____

13) $42 \times 76 =$ _____

14) $63 \times 34 =$ _____

15) $19 \times 31 =$ _____

16) $59 \times 71 =$ _____

✏️ **Estimate the sum or product by rounding each number to the nearest ten.**

17) $\begin{array}{r} 29 \\ \times\ 12 \\ \hline \end{array}$

18) $\begin{array}{r} 37 \\ \times\ 26 \\ \hline \end{array}$

19) $\begin{array}{r} 48 \\ +\ 82 \\ \hline \end{array}$

20) $\begin{array}{r} 65 \\ +44 \\ \hline \end{array}$

21) $\begin{array}{r} 37 \\ \times\ 14 \\ \hline \end{array}$

22) $\begin{array}{r} 71 \\ +\ 32 \\ \hline \end{array}$

WWW.MathNotion.Com

GED Subject Test – Mathematics

Adding and Subtracting Integers

✍ **Find each sum.**

1) $14 + (-6) =$

2) $(-13) + (-20) =$

3) $5 + (-28) =$

4) $50 + (-12) =$

5) $(-7) + (-15) + 3 =$

6) $30 + (-14) + 8 =$

7) $40 + (-10) + (-14) + 17 =$

8) $(-15) + (-20) + 13 + 35 =$

9) $40 + (-20) + (38 - 29) =$

10) $28 + (-12) + (30 - 12) =$

✍ **Find each difference.**

11) $(-18) - (-7) =$

12) $25 - (-14) =$

13) $(-20) - 36 =$

14) $34 - (-19) =$

15) $51 - (30 - 21) =$

16) $17 - (5) - (-24) =$

17) $(35 + 20) - (-46) =$

18) $48 - 16 - (-8) =$

19) $62 - (28 + 17) - (-15) =$

20) $58 - (-23) - (-31) =$

21) $19 - (-8) - (-13) =$

22) $(19 - 24) - (-14) =$

23) $27 - 33 - (-21) =$

24) $58 - (32 + 24) - (-9) =$

25) $36 - (-30) + (-17) =$

26) $27 - (-42) + (-31) =$

WWW.MathNotion.Com

GED Subject Test – Mathematics

Multiplying and Dividing Integers

✎ **Find each product.**

1) $(-9) \times (-5) =$

2) $(-3) \times 9 =$

3) $8 \times (-12) =$

4) $(-7) \times (-20) =$

5) $(-3) \times (-5) \times 6 =$

6) $(14 - 3) \times (-8) =$

7) $12 \times (-9) \times (-3) =$

8) $(140 + 10) \times (-2) =$

9) $10 \times (-12 + 8) \times 3 =$

10) $(-8) \times (-5) \times (-10) =$

✎ **Find each quotient.**

11) $42 \div (-7) =$

12) $(-48) \div (-6) =$

13) $(-40) \div (-8) =$

14) $54 \div (-2) =$

15) $152 \div 19 =$

16) $(-144) \div (-12) =$

17) $180 \div (-10) =$

18) $(-312) \div (-12) =$

19) $221 \div (-13) =$

20) $(-126) \div (6) =$

21) $(-161) \div (-7) =$

22) $-266 \div (-14) =$

23) $(-120) \div (-4) =$

24) $270 \div (-18) =$

25) $(-208) \div (-8) =$

26) $(135) \div (-15) =$

WWW.MathNotion.Com

GED Subject Test – Mathematics

Order of Operations

✎ **Evaluate each expression.**

1) $7 + (5 \times 4) =$

2) $14 - (3 \times 6) =$

3) $(19 \times 4) + 16 =$

4) $(16 - 7) - (8 \times 2) =$

5) $27 + (18 \div 3) =$

6) $(18 \times 8) \div 6 =$

7) $(32 \div 4) \times (-2) =$

8) $(9 \times 4) + (32 - 18) =$

9) $24 + (4 \times 3) + 7 =$

10) $(36 \times 3) \div (2 + 2) =$

11) $(-7) + (12 \times 3) + 11 =$

12) $(8 \times 5) - (24 \div 6) =$

13) $(7 \times 6 \div 3) - (12 + 9) =$

14) $(13 + 5 - 14) \times 3 - 2 =$

15) $(20 - 14 + 30) \times (64 \div 4) =$

16) $32 + (28 - (36 \div 9)) =$

17) $(7 + 6 - 4 - 7) + (15 \div 5) =$

18) $(85 - 20) + (20 - 18 + 7) =$

19) $(20 \times 2) + (14 \times 3) - 22 =$

20) $18 + 5 - (30 \times 3) + 20 =$

WWW.MathNotion.Com

GED Subject Test – Mathematics

Ordering Integers and Numbers

✏ **Order each set of integers from least to greatest.**

1) $8, -10, -5, -3, 4$ ___, ___, ___, ___, ___, ___

2) $-10, -18, 6, 14, 27$ ___, ___, ___, ___, ___, ___

3) $15, -8, -21, 21, -23$ ___, ___, ___, ___, ___, ___

4) $-14, -40, 23, -12, 47$ ___, ___, ___, ___, ___, ___

5) $59, -54, 32, -57, 36$ ___, ___, ___, ___, ___, ___

6) $68, 26, -19, 47, -34$ ___, ___, ___, ___, ___, ___

✏ **Order each set of integers from greatest to least.**

7) $18, 36, -16, -18, -10$ ___, ___, ___, ___, ___, ___

8) $27, 34, -12, -24, 94$ ___, ___, ___, ___, ___, ___

9) $50, -21, -13, 42, -2$ ___, ___, ___, ___, ___, ___

10) $37, 46, -20, -16, 86$ ___, ___, ___, ___, ___, ___

11) $-18, 88, -26, -59, 75$ ___, ___, ___, ___, ___, ___

12) $-65, -30, -25, 3, 14$ ___, ___, ___, ___, ___, ___

WWW.MathNotion.Com

GED Subject Test – Mathematics

Integers and Absolute Value

✎ **Write absolute value of each number.**

1) $|-2| =$

2) $|-27| =$

3) $|-20| =$

4) $|14| =$

5) $|6| =$

6) $|-55| =$

7) $|16| =$

8) $|2| =$

9) $|54| =$

10) $|-4| =$

11) $|-11|$

12) $|88| =$

13) $|0| =$

14) $|79| =$

15) $|-32| =$

16) $|-17| =$

17) $|42| =$

18) $|-46| =$

19) $|1| =$

20) $|-40| =$

✎ **Evaluate the value.**

21) $|-5| - \frac{|-21|}{7} =$

22) $14 - |3 - 15| - |-4| =$

23) $\frac{|-32|}{4} \times |-4| =$

24) $\frac{|7 \times (-3)|}{7} \times \frac{|-19|}{3} =$

25) $|4 \times (-5)| + \frac{|-40|}{5} =$

26) $\frac{|-45|}{9} \times \frac{|-24|}{12} =$

27) $|-12 + 8| \times \frac{|-7 \times 7|}{7}$

28) $\frac{|-11 \times 2|}{4} \times |-16| =$

GED Subject Test – Mathematics

Factoring Numbers

✎ **List all positive factors of each number.**

1) 9

2) 16

3) 24

4) 30

5) 26

6) 46

7) 20

8) 68

9) 28

10) 98

11) 14

12) 54

13) 55

14) 18

15) 63

16) 34

17) 50

18) 62

19) 95

20) 64

21) 70

22) 45

23) 22

24) 65

GED Subject Test – Mathematics

Greatest Common Factor

✎ **Find the GCF for each number pair.**

1) 6, 2

2) 4, 5

3) 3, 12

4) 7, 3

5) 5, 10

6) 8, 48

7) 6, 18

8) 9, 15

9) 12, 18

10) 4, 36

11) 6, 10

12) 28, 52

13) 25, 10

14) 22, 24

15) 9, 54

16) 8, 54

17) 42, 14

18) 16, 40

19) 9, 2, 3

20) 5, 15, 10

21) 7, 9, 2

22) 16, 64

23) 30, 48

24) 36, 63

Least Common Multiple

✎ **Find the LCM for each number pair.**

1) 6, 9

2) 15, 45

3) 16, 40

4) 12, 36

5) 18, 27

6) 14, 42

7) 6, 30

8) 8, 56

9) 7, 21

10) 8, 20

11) 15, 25

12) 7, 9

13) 4, 11

14) 8, 28

15) 28, 56

16) 40, 50

17) 12, 13

18) 22, 11

19) 36, 20

20) 15, 35

21) 18, 81

22) 30, 54

23) 18, 45

24) 75, 25

GED Subject Test – Mathematics

Answers of Worksheets

Rounding

1) 40	10) 80	19) 600	28) 7,000
2) 90	11) 60	20) 800	29) 9,000
3) 20	12) 90	21) 600	30) 23,000
4) 60	13) 200	22) 900	31) 45,000
5) 20	14) 400	23) 500	32) 17,000
6) 30	15) 800	24) 300	33) 53,000
7) 90	16) 100	25) 1,000	34) 85,000
8) 70	17) 300	26) 3,000	35) 71,000
9) 50	18) 400	27) 4,000	36) 27,000

Whole Number Addition and Subtraction

1) 1,898	7) 7,310	13) 640
2) 2,914	8) 9,116	14) 1,012
3) 1,507	9) 2,815	15) 3,095
4) 2,314	10) 3,092	16) 1,970
5) 3,317	11) 7,377	17) 970
6) 3,363	12) 3,371	18) 2,602

Whole Number Multiplication and Division

1) 546	7) 94	13) 300	19) 450
2) 1,050	8) 90	14) 300	20) 6
3) 560	9) 80	15) 80	21) 2
4) 440	10) 50	16) 70	22) 42
5) 1,210	11) 16	17) 80	
6) 2,400	12) 140	18) 250	

Rounding and Estimates

1) 30	7) 760	13) 3,200	19) 130
2) 90	8) 4,610	14) 1,800	20) 110
3) 120	9) 2,100	15) 600	21) 400
4) 150	10) 300	16) 4,200	22) 100
5) 620	11) 2,400	17) 300	
6) 1,230	12) 200	18) 1,200	

WWW.MathNotion.Com

GED Subject Test – Mathematics

Adding and Subtracting Integers

1) 8
2) −33
3) −23
4) 38
5) −19
6) 24
7) 33
8) 13
9) 29
10) 34
11) −11
12) 39
13) −56
14) 53
15) 42
16) 36
17) 101
18) 40
19) 32
20) 112
21) 40
22) 9
23) 15
24) 11
25) 49
26) 38

Multiplying and Dividing Integers

1) 45
2) −27
3) −96
4) 140
5) 90
6) −88
7) 324
8) −300
9) −120
10) −400
11) −6
12) 8
13) 5
14) −27
15) 8
16) 12
17) −18
18) 26
19) −17
20) −21
21) 23
22) 19
23) 30
24) −15
25) 26
26) −9

Order of Operations

1) 27
2) −4
3) 92
4) −7
5) 33
6) 24
7) −16
8) 50
9) 43
10) 27
11) 40
12) 36
13) −7
14) 10
15) 576
16) 56
17) 5
18) 74
19) 60
20) −47

Ordering Integers and Numbers

1) −10, −5, −3, 4, 8
2) −18, −10, 6, 14, 27
3) −23, −21, −8, 15, 21
4) −40, −14, −12, 23, 47
5) −57, −54, 32, 36, 59
6) −34, −19, 26, 47, 68
7) 36, 18, −10, −16, −18
8) 94, 34, 27, −12, −24
9) 50, 42, −2, −13, −21
10) 86, 46, 37, −16, −20
11) 88, 75, −18, −26, −59
12) 14, 3, −25, −30, −65

Integers and Absolute Value

1) 2
2) 27
3) 20
4) 14

GED Subject Test – Mathematics

5) 6	11) 11	17) 42	23) 32
6) 55	12) 88	18) 46	24) 19
7) 16	13) 0	19) 1	25) 28
8) 2	14) 79	20) 40	26) 10
9) 54	15) 32	21) 2	27) 28
10) 4	16) 17	22) −2	28) 88

Factoring Numbers

1) 1, 3, 9
2) 1, 2, 4, 8, 16
3) 1, 2, 3, 4, 6, 8, 12, 24
4) 1, 2, 3, 5, 6, 10, 15, 30
5) 1, 2, 13, 26
6) 1, 2, 23, 46
7) 1, 2, 4, 5, 10, 20
8) 1, 2, 4, 17, 34, 68
9) 1, 2, 4, 7, 14, 28
10) 1, 2, 7, 14, 49, 98
11) 1, 2, 7, 14
12) 1, 2, 3, 6, 9, 18, 27, 54
13) 1, 5, 11, 55
14) 1, 2, 3, 6, 9, 18
15) 1, 3, 7, 9, 21, 63
16) 1, 2, 17, 34
17) 1, 2, 5, 10, 25, 50
18) 1, 2, 31, 62
19) 1, 5, 19, 95
20) 1, 2, 4, 8, 16, 32, 64
21) 1, 2, 5, 7, 10, 14, 35, 70
22) 1, 3, 5, 9, 15, 45
23) 1, 2, 11, 22
24) 1, 5, 13, 65

Greatest Common Factor

1) 2	7) 6	13) 5	19) 1
2) 1	8) 3	14) 2	20) 5
3) 3	9) 6	15) 9	21) 1
4) 1	10) 4	16) 2	22) 16
5) 5	11) 2	17) 14	23) 6
6) 8	12) 4	18) 8	24) 9

Least Common Multiple

1) 18	7) 30	13) 44	19) 180
2) 45	8) 56	14) 56	20) 105
3) 80	9) 21	15) 56	21) 162
4) 36	10) 40	16) 200	22) 270
5) 54	11) 75	17) 156	23) 90
6) 42	12) 63	18) 22	24) 75

GED Subject Test – Mathematics

Chapter 2 :
Fractions and Decimals

Topics that you'll practice in this chapter:

- ✓ Simplifying Fractions
- ✓ Adding and Subtracting Fractions
- ✓ Multiplying and Dividing Fractions
- ✓ Adding and Subtract Mixed Numbers
- ✓ Multiplying and Dividing Mixed Numbers
- ✓ Adding and Subtracting Decimals
- ✓ Multiplying and Dividing Decimals
- ✓ Comparing Decimals
- ✓ Rounding Decimals

"A Man is like a fraction whose numerator is what he is and whose denominator is what he thinks of himself. The larger the denominator, the smaller the fraction." –Tolstoy

GED Subject Test – Mathematics

Simplifying Fractions

✎ **Simplify each fraction to its lowest terms.**

1) $\dfrac{5}{10} =$

2) $\dfrac{28}{35} =$

3) $\dfrac{27}{36} =$

4) $\dfrac{40}{80} =$

5) $\dfrac{14}{56} =$

6) $\dfrac{32}{48} =$

7) $\dfrac{52}{65} =$

8) $\dfrac{15}{60} =$

9) $\dfrac{80}{160} =$

10) $\dfrac{55}{77} =$

11) $\dfrac{28}{112} =$

12) $\dfrac{32}{64} =$

13) $\dfrac{63}{72} =$

14) $\dfrac{81}{90} =$

15) $\dfrac{35}{105} =$

16) $\dfrac{25}{70} =$

17) $\dfrac{80}{280} =$

18) $\dfrac{12}{81} =$

19) $\dfrac{36}{186} =$

20) $\dfrac{240}{540} =$

21) $\dfrac{70}{560} =$

✎ **Find the answer for each problem.**

22) Which of the following fractions equal to $\dfrac{3}{4}$? ____

    A. $\dfrac{60}{90}$ B. $\dfrac{43}{104}$ C. $\dfrac{48}{64}$ D. $\dfrac{150}{300}$

23) Which of the following fractions equal to $\dfrac{5}{8}$? ____

    A. $\dfrac{125}{200}$ B. $\dfrac{115}{200}$ C. $\dfrac{50}{100}$ D. $\dfrac{30}{90}$

24) Which of the following fractions equal to $\dfrac{3}{7}$? ____

    A. $\dfrac{58}{116}$ B. $\dfrac{54}{126}$ C. $\dfrac{270}{167}$ D. $\dfrac{42}{63}$

WWW.MathNotion.Com

GED Subject Test – Mathematics

Adding and Subtracting Fractions

✏ **Find the sum.**

1) $\frac{5}{9} + \frac{4}{9} =$

2) $\frac{1}{2} + \frac{1}{7} =$

3) $\frac{3}{8} + \frac{1}{4} =$

4) $\frac{3}{5} + \frac{1}{2} =$

5) $\frac{1}{4} + \frac{3}{5} =$

6) $\frac{7}{8} + \frac{3}{8} =$

7) $\frac{1}{2} + \frac{7}{10} =$

8) $\frac{2}{5} + \frac{2}{3} =$

9) $\frac{5}{7} + \frac{2}{3} =$

10) $\frac{7}{12} + \frac{3}{4} =$

11) $\frac{5}{6} + \frac{2}{5} =$

12) $\frac{1}{12} + \frac{2}{3} =$

✏ **Find the difference.**

13) $\frac{1}{3} - \frac{1}{6} =$

14) $\frac{3}{4} - \frac{1}{8} =$

15) $\frac{1}{2} - \frac{1}{3} =$

16) $\frac{1}{4} - \frac{1}{5} =$

17) $\frac{5}{8} - \frac{2}{3} =$

18) $\frac{1}{4} - \frac{1}{7} =$

19) $\frac{5}{6} - \frac{1}{9} =$

20) $\frac{3}{4} - \frac{1}{6} =$

21) $\frac{7}{8} - \frac{1}{12} =$

22) $\frac{8}{15} - \frac{3}{5} =$

23) $\frac{3}{12} - \frac{1}{14} =$

24) $\frac{10}{13} - \frac{7}{26} =$

25) $\frac{6}{7} - \frac{3}{4} =$

26) $\frac{4}{5} - \frac{1}{8} =$

27) $\frac{4}{7} - \frac{2}{35} =$

28) $\frac{9}{16} - \frac{2}{8} =$

29) $\frac{8}{9} - \frac{7}{18} =$

30) $\frac{1}{2} - \frac{4}{9} =$

WWW.MathNotion.Com

GED Subject Test – Mathematics

Multiplying and Dividing Fractions

✏️ **Find the value of each expression in lowest terms.**

1) $\frac{1}{5} \times \frac{15}{5} =$

2) $\frac{9}{12} \times \frac{4}{9} =$

3) $\frac{1}{16} \times \frac{8}{10} =$

4) $\frac{1}{24} \times \frac{8}{10} =$

5) $\frac{1}{5} \times \frac{1}{4} =$

6) $\frac{7}{9} \times \frac{1}{7} =$

7) $\frac{6}{7} \times \frac{1}{3} =$

8) $\frac{2}{8} \times \frac{2}{8} =$

9) $\frac{5}{8} \times \frac{3}{5} =$

10) $\frac{4}{7} \times \frac{1}{8} =$

11) $\frac{7}{15} \times \frac{5}{7} =$

12) $\frac{3}{10} \times \frac{5}{9} =$

✏️ **Find the value of each expression in lowest terms.**

13) $\frac{1}{4} \div \frac{1}{8} =$

14) $\frac{1}{10} \div \frac{1}{5} =$

15) $\frac{3}{4} \div \frac{1}{5} =$

16) $\frac{1}{3} \div \frac{5}{6} =$

17) $\frac{1}{7} \div \frac{8}{42} =$

18) $\frac{3}{4} \div \frac{1}{6} =$

19) $\frac{2}{7} \div \frac{7}{13} =$

20) $\frac{1}{24} \div \frac{3}{16} =$

21) $\frac{7}{12} \div \frac{5}{6} =$

22) $\frac{22}{18} \div \frac{11}{9} =$

23) $\frac{9}{35} \div \frac{3}{7} =$

24) $\frac{2}{7} \div \frac{8}{21} =$

25) $\frac{1}{9} \div \frac{2}{5} =$

26) $\frac{5}{12} \div \frac{3}{5} =$

27) $\frac{3}{20} \div \frac{1}{6} =$

28) $\frac{8}{20} \div \frac{3}{4} =$

29) $\frac{5}{6} \div \frac{2}{9} =$

30) $\frac{5}{11} \div \frac{3}{4} =$

WWW.MathNotion.Com

GED Subject Test – Mathematics

Adding and Subtracting Mixed Numbers

✍ **Find the sum.**

1) $3\frac{1}{3} + 2\frac{1}{6} =$

2) $4\frac{1}{2} + 3\frac{1}{2} =$

3) $3\frac{3}{8} + 1\frac{1}{8} =$

4) $2\frac{1}{4} + 2\frac{1}{3} =$

5) $3\frac{5}{6} + 2\frac{7}{12} =$

6) $5\frac{4}{15} + 3\frac{3}{5} =$

7) $2\frac{1}{3} + 4\frac{3}{7} =$

8) $3\frac{1}{2} + 4\frac{2}{5} =$

9) $5\frac{2}{5} + 6\frac{3}{7} =$

10) $8\frac{5}{16} + 6\frac{1}{12} =$

✍ **Find the difference.**

11) $3\frac{1}{4} - 1\frac{3}{4} =$

12) $6\frac{3}{5} - 4\frac{2}{5} =$

13) $4\frac{1}{3} - 3\frac{1}{9} =$

14) $7\frac{1}{7} - 5\frac{1}{2} =$

15) $5\frac{1}{3} - 2\frac{1}{12} =$

16) $8\frac{1}{5} - 4\frac{1}{3} =$

17) $9\frac{1}{4} - 6\frac{1}{8} =$

18) $11\frac{7}{15} - 8\frac{3}{5} =$

19) $14\frac{5}{6} - 11\frac{3}{5} =$

20) $18\frac{2}{7} - 14\frac{1}{5} =$

21) $9\frac{1}{3} - 4\frac{1}{4} =$

22) $6\frac{1}{8} - 4\frac{1}{16} =$

23) $19\frac{3}{8} - 15\frac{1}{3} =$

24) $11\frac{1}{9} - 8\frac{1}{8} =$

25) $17\frac{1}{7} - 11\frac{1}{5} =$

26) $16\frac{2}{9} - 9\frac{5}{7} =$

WWW.MathNotion.Com

GED Subject Test – Mathematics

Multiplying and Dividing Mixed Numbers

✎ **Find the product.**

1) $5\frac{1}{2} \times 2\frac{1}{4} =$

2) $5\frac{1}{3} \times 4\frac{1}{3} =$

3) $5\frac{3}{4} \times 6\frac{1}{4} =$

4) $3\frac{1}{3} \times 2\frac{3}{5} =$

5) $4\frac{8}{10} \times 1\frac{1}{24} =$

6) $6\frac{2}{7} \times 1\frac{1}{11} =$

7) $8\frac{2}{3} \times 3\frac{1}{2} =$

8) $3\frac{4}{7} \times 2\frac{1}{5} =$

9) $5\frac{2}{8} \times 4\frac{1}{6} =$

10) $7\frac{3}{3} \times 1\frac{3}{8} =$

✎ **Find the quotient.**

11) $2\frac{2}{5} \div 4\frac{1}{5} =$

12) $4\frac{1}{6} \div 3\frac{1}{3} =$

13) $6\frac{1}{3} \div 1\frac{1}{2} =$

14) $7\frac{1}{10} \div 2\frac{2}{5} =$

15) $3\frac{1}{3} \div 1\frac{1}{9} =$

16) $1\frac{1}{10} \div 4\frac{1}{2} =$

17) $1\frac{3}{16} \div 5\frac{1}{4} =$

18) $4\frac{1}{3} \div 4\frac{3}{4} =$

19) $9\frac{1}{3} \div 2\frac{1}{4} =$

20) $15\frac{1}{3} \div 5\frac{1}{2} =$

21) $4\frac{1}{6} \div 1\frac{1}{5} =$

22) $1\frac{1}{18} \div 1\frac{2}{9} =$

23) $4\frac{2}{7} \div 1\frac{3}{10} =$

24) $7\frac{1}{3} \div 2\frac{2}{11} =$

25) $8\frac{2}{5} \div 1\frac{1}{6} =$

26) $9\frac{1}{3} \div 2\frac{1}{7} =$

GED Subject Test – Mathematics

Adding and Subtracting Decimals

✎ **Add and subtract decimals.**

1) 35.19 − 24.28 = _____

4) 38.72 − 21.68 = _____

7) 86.09 − 35.14 = _____

2) 34.29 + 42.58 = _____

5) 57.39 + 26.54 = _____

8) 54.51 + 32.66 = _____

3) 61.20 + 33.75 = _____

6) 70.24 − 42.35 = _____

9) 114.21 − 88.69 = _____

✎ **Find the missing number.**

10) ___ + 2.8 = 5.4

11) 4.1 + ___ = 5.88

12) 6.45 + ___ = 8

13) 7.25 − ___ = 3.40

14) ___ − 2.35 = 4.25

15) ___ − 19.85 = 6.54

16) 22.15 + ___ = 28.95

17) ___ − 37.16 = 9.42

18) ___ + 24.50 = 34.19

19) 72.40 + ___ = 125.20

WWW.MathNotion.Com

GED Subject Test – Mathematics

Multiplying and Dividing Decimals

✏ **Find the product.**

1) $0.5 \times 0.6 =$

2) $3.3 \times 0.4 =$

3) $1.28 \times 0.5 =$

4) $0.35 \times 0.6 =$

5) $1.85 \times 0.6 =$

6) $0.24 \times 0.5 =$

7) $5.25 \times 1.4 =$

8) $18.5 \times 4.6 =$

9) $15.4 \times 6.8 =$

10) $19.5 \times 2.6 =$

11) $32.2 \times 1.5 =$

12) $78.4 \times 4.5 =$

✏ **Find the quotient.**

13) $1.85 \div 10 =$

14) $74.6 \div 100 =$

15) $3.6 \div 3 =$

16) $9.6 \div 0.4 =$

17) $15.5 \div 0.5 =$

18) $32.8 \div 0.2 =$

19) $22.15 \div 1,000 =$

20) $53.55 \div 0.7 =$

21) $322.2 \div 0.2 =$

22) $50.67 \div 0.18 =$

23) $77.4 \div 0.8 =$

24) $27.93 \div 0.03 =$

Comparing Decimals

✎ **Write the correct comparison symbol (>, < or =).**

1) 0.70 ☐ 0.070

2) 0.049 ☐ 0.49

3) 5.090 ☐ 5.09

4) 2.57 ☐ 2.05

5) 9.03 ☐ 0.930

6) 6.06 ☐ 6.6

7) 7.02 ☐ 7.020

8) 3.04 ☐ 3.2

9) 3.61 ☐ 3.245

10) 0.986 ☐ 0.0986

11) 17.24 ☐ 17.240

12) 0.759 ☐ 0.81

13) 9.040 ☐ 9.40

14) 5.73 ☐ 5.213

15) 9.44 ☐ 9.404

16) 7.17 ☐ 7.170

17) 4.85 ☐ 4.085

18) 9.041 ☐ 9.40

19) 3.033 ☐ 3.030

20) 4.97 ☐ 4.970

GED Subject Test – Mathematics

Rounding Decimals

✏ Round each decimal to the nearest whole number.

1) 28.12   3) 16.22   5) 7.95

2) 6.9   4) 8.5   6) 52.7

✏ Round each decimal to the nearest tenth.

7) 31.761   9) 94.729   11) 13.219

8) 14.421   10) 77.89   12) 59.89

✏ Round each decimal to the nearest hundredth.

13) 8.428   15) 55.3786   17) 62.241

14) 23.812 16) 231.912   18) 19.447

✏ Round each decimal to the nearest thousandth.

19) 15.54324   21) 243.8652   23) 67.1983

20) 34.62586   22) 80.4529   24) 72.36788

GED Subject Test – Mathematics

Answers of Worksheets

Simplifying Fractions

1) $\frac{1}{2}$
2) $\frac{4}{5}$
3) $\frac{3}{4}$
4) $\frac{1}{2}$
5) $\frac{1}{4}$
6) $\frac{2}{3}$
7) $\frac{4}{5}$
8) $\frac{1}{4}$
9) $\frac{1}{2}$
10) $\frac{5}{7}$
11) $\frac{1}{4}$
12) $\frac{1}{2}$
13) $\frac{7}{8}$
14) $\frac{9}{10}$
15) $\frac{1}{3}$
16) $\frac{5}{14}$
17) $\frac{2}{7}$
18) $\frac{4}{27}$
19) $\frac{6}{31}$
20) $\frac{4}{9}$
21) $\frac{1}{8}$
22) C
23) A
24) B

Adding and Subtracting Fractions

1) $\frac{9}{9} = 1$
2) $\frac{9}{14}$
3) $\frac{5}{8}$
4) $1\frac{1}{10}$
5) $\frac{17}{20}$
6) $1\frac{1}{4}$
7) $1\frac{1}{5}$
8) $1\frac{1}{15}$
9) $1\frac{8}{21}$
10) $1\frac{1}{3}$
11) $1\frac{7}{30}$
12) $\frac{3}{4}$
13) $\frac{1}{6}$
14) $\frac{5}{8}$
15) $\frac{1}{6}$
16) $\frac{1}{20}$
17) $-\frac{1}{24}$
18) $\frac{3}{28}$
19) $\frac{13}{18}$
20) $\frac{7}{12}$
21) $\frac{19}{24}$
22) $-\frac{1}{15}$
23) $\frac{5}{28}$
24) $\frac{1}{2}$
25) $\frac{3}{28}$
26) $\frac{27}{40}$
27) $\frac{18}{35}$
28) $\frac{5}{16}$
29) $\frac{1}{2}$
30) $\frac{1}{18}$

Multiplying and Dividing Fractions

1) $\frac{3}{5}$
2) $\frac{1}{3}$
3) $\frac{1}{20}$
4) $\frac{1}{30}$
5) $\frac{1}{20}$
6) $\frac{1}{9}$
7) $\frac{2}{7}$
8) $\frac{1}{16}$
9) $\frac{3}{8}$
10) $\frac{1}{14}$
11) $\frac{1}{3}$
12) $\frac{1}{6}$
13) 2
14) $\frac{1}{2}$

WWW.MathNotion.Com

GED Subject Test – Mathematics

15) $3\frac{3}{4}$ 19) $\frac{26}{49}$ 23) $\frac{3}{5}$ 27) $\frac{9}{10}$

16) $\frac{2}{5}$ 20) $\frac{2}{9}$ 24) $\frac{3}{4}$ 28) $\frac{8}{15}$

17) $\frac{3}{4}$ 21) $\frac{7}{10}$ 25) $\frac{5}{18}$ 29) $3\frac{3}{4}$

18) $4\frac{1}{2}$ 22) 1   26) $\frac{25}{36}$ 30) $\frac{20}{33}$

Adding and Subtracting Mixed Numbers

1) $5\frac{1}{2}$ 8) $7\frac{9}{10}$ 15) $3\frac{1}{4}$ 22) $2\frac{1}{16}$

2) 8   9) $11\frac{29}{35}$ 16) $3\frac{13}{15}$ 23) $4\frac{1}{24}$

3) $4\frac{1}{2}$ 10) $14\frac{19}{48}$ 17) $3\frac{1}{8}$ 24) $2\frac{71}{72}$

4) $4\frac{7}{12}$ 11) $1\frac{1}{2}$ 18) $2\frac{13}{15}$ 25) $5\frac{33}{35}$

5) $6\frac{5}{12}$ 12) $2\frac{1}{5}$ 19) $3\frac{7}{30}$ 26) $6\frac{32}{63}$

6) $8\frac{13}{15}$ 13) $1\frac{2}{9}$ 20) $4\frac{3}{35}$

7) $6\frac{16}{21}$ 14) $1\frac{9}{14}$ 21) $5\frac{1}{12}$

Multiplying and Dividing Mixed Numbers

1) $12\frac{3}{8}$ 10) 11   19) $4\frac{4}{27}$

2) $23\frac{1}{9}$ 11) $\frac{4}{7}$ 20) $2\frac{26}{33}$

3) $35\frac{15}{16}$ 12) $1\frac{1}{4}$ 21) $3\frac{17}{36}$

4) $8\frac{2}{3}$ 13) $4\frac{2}{9}$ 22) $\frac{19}{22}$

5) 5   14) $2\frac{23}{24}$ 23) $3\frac{27}{91}$

6) $6\frac{6}{7}$ 15) 3   24) $3\frac{13}{36}$

7) $30\frac{1}{3}$ 16) $\frac{11}{45}$ 25) $7\frac{1}{5}$

8) $7\frac{6}{7}$ 17) $\frac{19}{84}$ 26) $4\frac{16}{45}$

9) $21\frac{7}{8}$ 18) $\frac{52}{57}$

Adding and Subtracting Decimals

1) 10.91   2) 76.87   3) 94.95   4) 17.04

GED Subject Test – Mathematics

5) 83.93	9) 25.52	13) 3.85	17) 46.58
6) 27.89	10) 2.6	14) 6.6	18) 9.69
7) 50.95	11) 1.78	15) 26.39	19) 52.8
8) 87.17	12) 1.55	16) 6.8	

Multiplying and Dividing Decimals

1) 0.3	7) 7.35	13) 0.185	19) 0.02215
2) 1.32	8) 85.1	14) 0.746	20) 76.5
3) 0.64	9) 104.72	15) 1.2	21) 1,611
4) 0.21	10) 50.7	16) 24	22) 281.5
5) 1.11	11) 48.3	17) 31	23) 96.75
6) 0.12	12) 352.8	18) 164	24) 931

Comparing Decimals

1) >	6) <	11) =	16) =
2) <	7) =	12) <	17) >
3) =	8) <	13) <	18) <
4) >	9) >	14) >	19) >
5) >	10) >	15) >	20) =

Rounding Decimals

1) 28	9) 94.7	17) 62.24	
2) 7	10) 77.9	18) 19.45	
3) 16	11) 13.2	19) 15.543	
4) 9	12) 59.9	20) 34.626	
5) 8	13) 8.43	21) 243.865	
6) 53	14) 23.81	22) 80.453	
7) 31.8	15) 55.38	23) 67.198	
8) 14.4	16) 231.91	24) 72.368	

GED Subject Test – Mathematics

Chapter 3 :
Proportions, Ratios, and Percent

Topics that you'll practice in this chapter:

- ✓ Simplifying Ratios
- ✓ Proportional Ratios
- ✓ Similarity and Ratios
- ✓ Ratio and Rates Word Problems
- ✓ Percentage Calculations
- ✓ Percent Problems
- ✓ Discount, Tax and Tip
- ✓ Percent of Change
- ✓ Simple Interest

Without mathematics, there's nothing you can do. Everything around you is mathematics. Everything around you is numbers." – *Shakuntala Devi*

GED Subject Test – Mathematics

Simplifying Ratios

✏ Reduce each ratio.

1) 15: 20 = ___:___
2) 7: 70 = ___:___
3) 16: 28 = ___:___
4) 7: 21 = ___:___
5) 4: 40 = ___:___
6) 6: 48 = ___:___
7) 16: 64 = ___:___
8) 10: 25 = ___:___

9) 8: 48 = ___:___
10) 49: 63 = ___:___
11) 18: 27 = ___:___
12) 35: 10 = ___:___
13) 90: 9 = ___:___
14) 24: 32 = ___:___
15) 7: 56 = ___:___
16) 45: 63 = ___:___

17) 56: 72 = ___:___
18) 26: 13 = ___:___
19) 15: 45 = ___:___
20) 28: 4 = ___:___
21) 24: 48 = ___:___
22) 30: 24 = ___:___
23) 70: 140 = ___:___
24) 6: 180 = ___:___

✏ Write each ratio as a fraction in simplest form.

25) 6: 12 =
26) 30: 50 =
27) 15: 35 =
28) 9: 27 =
29) 8: 24 =
30) 18: 84 =
31) 7: 14 =

32) 7: 35 =
33) 40: 96 =
34) 12: 54 =
35) 44: 52 =
36) 12: 27 =
37) 15: 180 =
38) 39: 143 =

39) 20: 300 =
40) 30: 120 =
41) 56: 42 =
42) 26: 130 =
43) 66: 123 =
44) 70: 630 =
45) 75: 125 =

GED Subject Test – Mathematics

Proportional Ratios

✎ Fill in the blanks; Calculate each proportion.

1) $3:8 = \_\_ : 48$ 7) $20:3 = \_\_ : 15$

2) $2:5 = 20:\_\_$ 8) $1:3 = \_\_ : 75$

3) $1:9 = \_\_ : 81$ 9) $7:6 = \_\_ : 60$

4) $6:7 = 12:\_\_$ 10) $8:5 = \_\_ : 45$

5) $9:2 = 63:\_\_$ 11) $3:10 = 60:\_\_$

6) $8:7 = \_\_ : 49$ 12) $6:11 = 42:\_\_$

✎ State if each pair of ratios form a proportion.

13) $\frac{3}{20}$ and $\frac{9}{60}$ 17) $\frac{1}{9}$ and $\frac{12}{81}$ 21) $\frac{6}{19}$ and $\frac{30}{85}$

14) $\frac{1}{7}$ and $\frac{6}{42}$ 18) $\frac{7}{8}$ and $\frac{21}{28}$ 22) $\frac{5}{9}$ and $\frac{40}{81}$

15) $\frac{3}{7}$ and $\frac{24}{56}$ 19) $\frac{9}{13}$ and $\frac{27}{39}$ 23) $\frac{9}{14}$ and $\frac{108}{168}$

16) $\frac{4}{9}$ and $\frac{12}{18}$ 20) $\frac{1}{8}$ and $\frac{8}{64}$ 24) $\frac{15}{23}$ and $\frac{360}{552}$

✎ Calculate each proportion.

25) $\frac{20}{25} = \frac{32}{x}, x = \_\_\_$ 29) $\frac{7}{9} = \frac{x}{81}, x = \_\_\_$ 33) $\frac{5}{8} = \frac{x}{88}, x = \_\_\_$

26) $\frac{1}{8} = \frac{32}{x}, x = \_\_\_$ 30) $\frac{1}{5} = \frac{13}{x}, x = \_\_\_$ 34) $\frac{4}{15} = \frac{x}{240}, x = \_\_\_$

27) $\frac{15}{5} = \frac{21}{x}, x = \_\_\_$ 31) $\frac{9}{5} = \frac{36}{x}, x = \_\_\_$ 35) $\frac{9}{19} = \frac{x}{266}, x = \_\_\_$

28) $\frac{1}{7} = \frac{x}{294}, x = \_\_\_$ 32) $\frac{6}{13} = \frac{48}{x}, x = \_\_\_$ 36) $\frac{7}{15} = \frac{x}{270}, x = \_\_\_$

WWW.MathNotion.Com

GED Subject Test – Mathematics

Similarity and Ratios

✎ **Each pair of figures is similar. Find the missing side.**

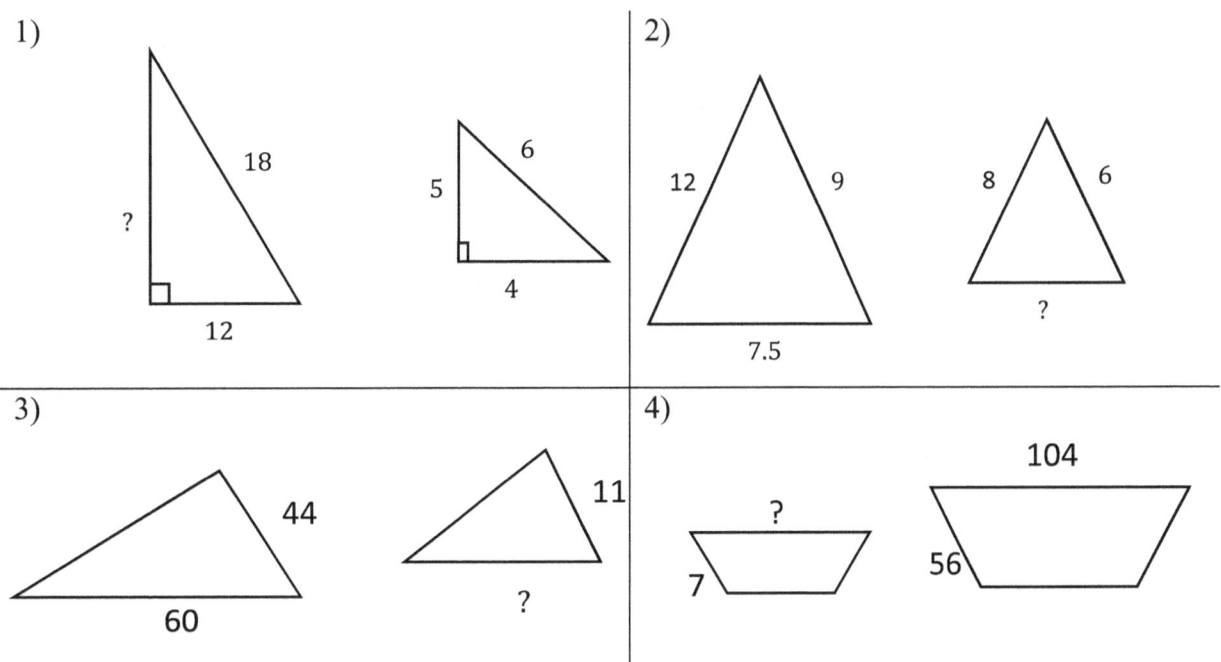

✎ **Calculate.**

5) Two rectangles are similar. The first is 24 feet wide and 120 feet long. The second is 30 feet wide. What is the length of the second rectangle? _____

6) Two rectangles are similar. One is 5 meters by 36 meters. The longer side of the second rectangle is 90 meters. What is the other side of the second rectangle? _____

7) A building casts a shadow 25 ft long. At the same time a girl 10 ft tall casts a shadow 5 ft long. How tall is the building? _____

8) The scale of a map of Texas is 4 inches: 32 miles. If you measure the distance from Dallas to Martin County as 38.4 inches, approximately how far is Martin County from Dallas? _____

GED Subject Test – Mathematics

Ratio and Rates Word Problems

✎ Find the answer for each word problem.

1) Mason has 24 red cards and 36 green cards. What is the ratio of Mason's red cards to his green cards? _____

2) In a party, 45 soft drinks are required for every 54 guests. If there are 378 guests, how many soft drinks is required? _____

3) In Mason's class, 42 of the students are tall and 24 are short. In Michael's class 84 students are tall and 48 students are short. Which class has a higher ratio of tall to short students? _____

4) The price of 5 apples at the Quick Market is $4.6. The price of 7 of the same apples at Walmart is $5.95. Which place is the better buy? _____

5) The bakers at a Bakery can make 90 bagels in 3 hours. How many bagels can they bake in 24 hours? What is that rate per hour? _____

6) You can buy 5 cans of green beans at a supermarket for $5.75. How much does it cost to buy 45 cans of green beans? _____

7) The ratio of boys to girls in a class is 4: 7. If there are 32 boys in the class, how many girls are in that class? _____

8) The ratio of red marbles to blue marbles in a bag is 3: 7. If there are 50 marbles in the bag, how many of the marbles are red? _____

WWW.MathNotion.Com

GED Subject Test – Mathematics

Percentage Calculations

✎ **Calculate the given percent of each value.**

1) 3% of 60 = ____
2) 20% of 32 = ____
3) 4% of 72 = ____
4) 16% of 32 = ____
5) 25% of 124 = ____
6) 35% of 56 = ____
7) 15% of 20 = ____
8) 14% of 150 = ____
9) 80% of 50 = ____
10) 12% of 115 = ____
11) 72% of 250 = ____
12) 52% of 500 = ____
13) 70% of 400 = ____
14) 27% of 145 = ____
15) 90% of 64 = ____
16) 60% of 55 = ____
17) 22% of 210 = ____
18) 8% of 235 = ____

✎ **Calculate the percent of each given value.**

19) ____% of 25 = 5
20) ____% of 40 = 20
21) ____% of 25 = 2
22) ____% of 50 = 16
23) ____% of 250 = 5
24) ____% of 40 = 32
25) ____% of 125 = 20
26) ____% of 700 = 49
27) ____% of 350 = 49
28) ____% of 500 = 210

✎ **Calculate each percent problem.**

29) A Cinema has 250 seats. 60 seats were sold for the current movie. What percent of seats are empty? _____ %

30) There are 68 boys and 92 girls in a class. 75% of the students in the class take the bus to school. How many students do not take the bus to school? _____

GED Subject Test – Mathematics

Percent Problems

✎ **Calculate each problem.**

1) 9 is what percent of 45? ___%

2) 60 is what percent of 120? ___%

3) 10 is what percent of 200? ___%

4) 15 is what percent of 125? ___%

5) 10 is what percent of 400? ___%

6) 66 is what percent of 55? ___%

7) 40 is what percent of 160? ___%

8) 40 is what percent of 50? ___%

9) 120 is what percent of 800? ___%

10) 78 is what percent of 120? ___%

11) 36 is what percent of 144? ___%

12) 17 is what percent of 85? ___%

13) 90 is what percent of 900? ___%

14) 36 is what percent of 16? ___%

15) 63 is what percent of 14? ___%

16) 18 is what percent of 60? ___%

17) 126 is what percent of 200? ___%

18) 232 is what percent of 40? ___%

✎ **Calculate each percent word problem.**

19) There are 40 employees in a company. On a certain day, 25 were present. What percent showed up for work? ____%

20) A metal bar weighs 60 ounces. 25% of the bar is gold. How many ounces of gold are in the bar? _____

21) A crew is made up of 12 women; the rest are men. If 15% of the crew are women, how many people are in the crew? _____

22) There are 40 students in a class and 8 of them are girls. What percent are boys? ____%

23) The Royals softball team played 400 games and won 280 of them. What percent of the games did they lose? ____%

WWW.MathNotion.Com

GED Subject Test – Mathematics

Discount, Tax and Tip

✎ **Find the selling price of each item.**

1) Original price of a computer: $420
   Tax: 8% Selling price: $_____

2) Original price of a laptop: $280
   Tax: 4% Selling price: $_____

3) Original price of a sofa: $820
   Tax: 5% Selling price: $_____

4) Original price of a car: $15,800
   Tax: 3.6%   Selling price: $_____

5) Original price of a Table: $250
   Tax: 9% Selling price: $_____

6) Original price of a house: $630,000
   Tax: 1.8%   Selling price: $_____

7) Original price of a tablet: $450
   Discount: 30%   Selling price: $____

8) Original price of a chair: $390
   Discount: 8%   Selling price: $____

9) Original price of a book: $75
   Discount: 42%   Selling price: $____

10) Original price of a cellphone: $820
    Discount: 23%   Selling price: $____

11) Food bill: $45
    Tip: 15%   Price: $_____

12) Food bill: $32
    Tipp: 20%   Price: $_____

13) Food bill: $90
    Tip: 35%   Price: $_____

14) Food bill: $42
    Tipp: 12%   Price: $_____

✎ **Find the answer for each word problem.**

15) Nicolas hired a moving company. The company charged $500 for its services, and Nicolas gives the movers a 40% tip. How much does Nicolas tip the movers? $_____

16) Mason has lunch at a restaurant and the cost of his meal is $90. Mason wants to leave a 25% tip. What is Mason's total bill including tip? $_____

17) The sales tax in Texas is 19.80% and an item costs $350. How much is the tax? $_____

18) The price of a table at Best Buy is $680. If the sales tax is 5%, what is the final price of the table including tax? $_____

WWW.MathNotion.Com

GED Subject Test – Mathematics

Percent of Change

✎ **Find each percent of change.**

1) From 150 to 450. ___ %

2) From 50 ft to 250 ft. ___ %

3) From $60 to $360. ___ %

4) From 60 cm to 180 cm. ___ %

5) From 15 to 45. ___ %

6) From 80 to 16. ___ %

7) From 120 to 360. ___ %

8) From 900 to 450. ___ %

9) From 1,000 to 200. ___ %

10) From 144 to 36. ___ %

✎ **Calculate each percent of change word problem.**

11) Bob got a raise, and his hourly wage increased from $42 to $63. What is the percent increase? ___ %

12) The price of a pair of shoes increases from $50 to $61. What is the percent increase? ___ %

13) At a coffee shop, the price of a cup of coffee increased from $4.80 to $5.76. What is the percent increase in the cost of the coffee? ___ %

14) 51 cm are cut from 85 cm board. What is the percent decrease in length? ___ %

15) In a class, the number of students has been increased from 54 to 81. What is the percent increase? ___ %

16) The price of gasoline rises from $24.40 to $30.50 in one month. By what percent did the gas price rise? ___ %

17) A shirt was originally priced at $38. It went on sale for $24.70. What was the percent that the shirt was discounted? ___ %

GED Subject Test – Mathematics

Simple Interest

✎ **Determine the simple interest for these loans.**

1) $480 at 11% for 3 years. $ _____

2) $4,200 at 7% for 4 years. $ _____

3) $2,500 at 20% for 3 years. $ _____

4) $6,800 at 3.9% for 4 months. $ ____

5) $800 at 6% for 7 months. $ _____

6) $36,000 at 4.2% for 6 years. $ _____

7) $6,500 at 7% for 4 years. $ _____

8) $850 at 9.5% for 2 years. $ _____

9) $1,200 at 5.8% for 9 months. $ ____

10) $3,000 at 4.5% for 7 years. $ _____

✎ **Calculate each simple interest word problem.**

11) A new car, valued at $22,000, depreciates at 8.5% per year. What is the value of the car one year after purchase? $_____

12) Sara puts $9,000 into an investment yielding 6% annual simple interest; she left the money in for three years. How much interest does Sara get at the end of those three years? $_____

13) A bank is offering 12% simple interest on a savings account. If you deposit $16,400, how much interest will you earn in two years? $_____

14) $720 interest is earned on a principal of $6,000 at a simple interest rate of 4% interest per year. For how many years was the principal invested? _____

15) In how many years will $2,200 yield an interest of $440 at 4% simple interest? _____

16) Jim invested $8,000 in a bond at a yearly rate of 4.5%. He earned $1,440 in interest. How long was the money invested? _____

GED Subject Test – Mathematics

Answers of Worksheets

Simplifying Ratios

1) 3 : 4	14) 3 : 4	26) $\frac{3}{5}$	36) $\frac{4}{9}$
2) 1 : 10	15) 1 : 8	27) $\frac{3}{7}$	37) $\frac{1}{12}$
3) 4 : 7	16) 5 : 7	28) $\frac{1}{3}$	38) $\frac{3}{11}$
4) 1 : 3	17) 7 : 9	29) $\frac{1}{3}$	39) $\frac{1}{15}$
5) 1 : 10	18) 2 : 1	30) $\frac{3}{14}$	40) $\frac{1}{4}$
6) 1 : 8	19) 1 : 3	31) $\frac{1}{2}$	41) $\frac{4}{3}$
7) 2 : 8	20) 7 : 1	32) $\frac{1}{5}$	42) $\frac{1}{5}$
8) 2 : 5	21) 1 : 2	33) $\frac{5}{12}$	43) $\frac{22}{41}$
9) 1 : 6	22) 5 : 4	34) $\frac{2}{9}$	44) $\frac{1}{9}$
10) 7 : 9	23) 1 : 2	35) $\frac{11}{13}$	45) $\frac{3}{5}$
11) 2 : 3	24) 1 : 30		
12) 7 : 2	25) $\frac{1}{2}$		
13) 10 : 1			

Proportional Ratios

1) 18	10) 72	19) Yes	28) 42
2) 50	11) 200	20) Yes	29) 63
3) 9	12) 77	21) No	30) 65
4) 14	13) Yes	22) No	31) 20
5) 14	14) Yes	23) Yes	32) 104
6) 56	15) Yes	24) Yes	33) 55
7) 100	16) No	25) 40	34) 64
8) 25	17) No	26) 256	35) 126
9) 70	18) No	27) 7	36) 126

Similarity and ratios

1) 15	4) 13	7) 50 feet
2) 5	5) 150 feet	8) 307.2 miles
3) 15	6) 12.5 meters	

Ratio and Rates Word Problems

1) 2 : 3   2) 315

WWW.MathNotion.Com

GED Subject Test – Mathematics

3) The ratio for both classes is 7 to 4.   6) $51.75
4) Walmart is a better buy.   7) 56
5) 720, the rate is 30 per hour.   8) 15

Percentage Calculations

1) 1.8   11) 180   21) 8%
2) 6.4   12) 260   22) 32%
3) 2.88   13) 280   23) 2%
4) 5.12   14) 39.15   24) 80%
5) 31   15) 57.6   25) 16%
6) 19.6   16) 33   26) 7%
7) 3   17) 46.2   27) 14%
8) 21   18) 18.8   28) 42%
9) 40   19) 20%   29) 76%
10) 13.8   20) 50%   30) 40

Percent Problems

1) 20%   9) 15%   17) 63%
2) 50%   10) 65%   18) 580%
3) 5%   11) 25%   19) 62.5%
4) 12%   12) 20%   20) 15 ounces
5) 2.5%   13) 10%   21) 80
6) 120%   14) 225%   22) 80%
7) 25%   15) 450%   23) 30%
8) 80%   16) 30%

Discount, Tax and Tip

1) $453.60   7) $315.00   13) $121.50
2) $291.20   8) $358.80   14) $47.04
3) $861.00   9) $43.50   15) $200.00
4) $16,368.80   10) $631.40   16) $112.50
5) $272.50   11) $51.75   17) $69.30
6) $641,340   12) $38.40   18) $714.00

WWW.MathNotion.Com

GED Subject Test – Mathematics

Percent of Change

1) 200%
2) 400%
3) 500%
4) 200%
5) 200%
6) 80%
7) 200%
8) 50%
9) 80%
10) 75%
11) 50%
12) 22%
13) 20%
14) 60%
15) 50%
16) 25%
17) 35%

Simple Interest

1) $158.40
2) $1,176.00
3) $1,500.00
4) $88.40
5) $28.00
6) $9,072.00
7) $1,820.00
8) $161.50
9) $52.20
10) $945.00
11) $20,130.00
12) $1,620.00
13) $3,936.00
14) 3 years
15) 5 years
16) 4 years

GED Subject Test – Mathematics

Chapter 4 :
Exponents and Radicals Expressions

Topics that you'll practice in this chapter:

- ✓ Multiplication Property of Exponents
- ✓ Zero and Negative Exponents
- ✓ Division Property of Exponents
- ✓ Powers of Products and Quotients
- ✓ Negative Exponents and Negative Bases
- ✓ Scientific Notation
- ✓ Square Roots
- ✓ Simplifying Radical Expressions

Mathematics is no more computation than typing is literature.
– *John Allen Paulos*

GED Subject Test – Mathematics

Multiplication Property of Exponents

✏ Simplify and write the answer in exponential form.

1) $4 \times 4^5 =$

2) $8^4 \times 8 =$

3) $7^3 \times 7^3 =$

4) $9^2 \times 9^2 =$

5) $2^2 \times 2^4 \times 2 =$

6) $5 \times 5^3 \times 5^3 =$

7) $4^3 \times 4^2 \times 4 \times 4 =$

8) $5x \times x =$

9) $x^3 \times x^3 =$

10) $x^7 \times x^2 =$

11) $x^4 \times x^3 \times x^2 =$

12) $10x \times 3x =$

13) $4x^3 \times 4x^3 =$

14) $7x^3 \times x =$

15) $3x^2 \times 4x^2 \times x^2 =$

16) $5x^4 \times x^4 =$

17) $2x^8 \times 2x =$

18) $6x \times x^5 =$

19) $4x^2 \times 6x^6 =$

20) $5yx^3 \times 4x =$

21) $7x^3 \times y^5 x^7 =$

22) $y^2 x^3 \times y^5 x^4 =$

23) $3x^5 \times 4x^3 y^4 =$

24) $4x^4 \times 9x^2 y^5 =$

25) $5x^3 y^4 \times 6x^8 y^2 =$

26) $8x^3 y^6 \times 4xy^3 =$

27) $2xy^5 \times 6x^3 y^3 =$

28) $4x^5 y^2 \times 4x^2 y^8 =$

29) $7x \times 3y^8 x^2 \times y^5 =$

30) $x^3 \times 2y^3 x^4 \times 2y =$

31) $3yx^4 \times 3y^4 x \times 3xy^3 =$

32) $6y^3 \times 2y^2 x^4 \times 10yx^5 =$

WWW.MathNotion.Com

GED Subject Test – Mathematics

Zero and Negative Exponents

✎ **Evaluate the following expressions.**

1) $1^{-5} =$

2) $4^{-1} =$

3) $0^{10} =$

4) $1^{15} =$

5) $5^{-2} =$

6) $3^{-3} =$

7) $9^{-1} =$

8) $10^{-2} =$

9) $12^{-2} =$

10) $2^{-5} =$

11) $3^{-4} =$

12) $2^{-4} =$

13) $6^{-3} =$

14) $10^{-3} =$

15) $30^{-1} =$

16) $15^{-2} =$

17) $4^{-3} =$

18) $2^{-7} =$

19) $5^{-3} =$

20) $4^{-4} =$

21) $3^{-5} =$

22) $10^{-4} =$

23) $2^{-10} =$

24) $8^{-3} =$

25) $20^{-2} =$

26) $14^{-2} =$

27) $9^{-3} =$

28) $100^{-2} =$

29) $5^{-4} =$

30) $4^{-6} =$

31) $\left(\frac{1}{4}\right)^{-3} =$

32) $\left(\frac{1}{6}\right)^{-2} =$

33) $\left(\frac{1}{7}\right)^{-2} =$

34) $\left(\frac{2}{3}\right)^{-3} =$

35) $\left(\frac{1}{13}\right)^{-2} =$

36) $\left(\frac{7}{12}\right)^{-2} =$

37) $\left(\frac{1}{6}\right)^{-3} =$

38) $\left(\frac{1}{300}\right)^{-2} =$

39) $\left(\frac{2}{9}\right)^{-2} =$

40) $\left(\frac{7}{5}\right)^{-1} =$

41) $\left(\frac{13}{23}\right)^{0} =$

42) $\left(\frac{1}{4}\right)^{-5} =$

WWW.MathNotion.Com

GED Subject Test – Mathematics

Division Property of Exponents

✎ **Simplify.**

1) $\dfrac{5^6}{5^7} =$

2) $\dfrac{8^8}{8^6} =$

3) $\dfrac{4^5}{4} =$

4) $\dfrac{3}{3^5} =$

5) $\dfrac{x}{x^6} =$

6) $\dfrac{3 \times 3^2}{3^2 \times 3^5} =$

7) $\dfrac{9^4}{9^2} =$

8) $\dfrac{10 \times 10^9}{10^2 \times 10^7} =$

9) $\dfrac{7^5 \times 7^7}{7^4 \times 7^8} =$

10) $\dfrac{15x}{30x^6} =$

11) $\dfrac{3x^9}{4x^4} =$

12) $\dfrac{15x^8}{10x^9} =$

13) $\dfrac{42x^5}{6y^9} =$

14) $\dfrac{36y^8}{4x^4y^5} =$

15) $\dfrac{2x^7}{9x} =$

16) $\dfrac{49x^8y^6}{7x^9} =$

17) $\dfrac{48x^2}{24x^6y^{12}} =$

18) $\dfrac{30yx^5}{6yx^7} =$

19) $\dfrac{19x^7y}{38x^{12}y^4} =$

20) $\dfrac{9x^8}{63x^8} =$

21) $\dfrac{9x^{-9}}{4x^{-3}} =$

WWW.MathNotion.Com

GED Subject Test – Mathematics

Powers of Products and Quotients

✎ Simplify.

1) $(4^3)^2 =$

2) $(2^3)^4 =$

3) $(2 \times 2^3)^2 =$

4) $(5 \times 5^5)^6 =$

5) $(19^4 \times 19^2)^3 =$

6) $(2^3 \times 2^4)^4 =$

7) $(5 \times 5^2)^2 =$

8) $(4^4)^4 =$

9) $(8x^5)^2 =$

10) $(3x^2y^4)^4 =$

11) $(7x^5y^2)^2 =$

12) $(5x^4y^4)^3 =$

13) $(2x^3y^3)^5 =$

14) $(10x^3y^4)^3 =$

15) $(13y^3y)^2 =$

16) $(5x^6x^4)^2 =$

17) $(6x^7y^6)^3 =$

18) $(12x^5x^7)^2 =$

19) $(2x^4 \times 2x)^4 =$

20) $(2x^4y^3)^5 =$

21) $(15x^7y^2)^2 =$

22) $(8x^3y^5)^3 =$

23) $(3x \times 2y^2)^4 =$

24) $\left(\dfrac{4x}{x^5}\right)^2 =$

25) $\left(\dfrac{x^4y^5}{x^3y^5}\right)^9 =$

26) $\left(\dfrac{36xy}{6x^5}\right)^3 =$

27) $\left(\dfrac{x^7}{x^8y^2}\right)^6 =$

28) $\left(\dfrac{xy^4}{x^3y^6}\right)^{-3} =$

29) $\left(\dfrac{5xy^8}{x^3}\right)^2 =$

30) $\left(\dfrac{xy^6}{2xy^3}\right)^{-4} =$

WWW.MathNotion.Com

GED Subject Test – Mathematics

Negative Exponents and Negative Bases

✎ **Simplify.**

1) $-9^{-1} =$

2) $-9^{-2} =$

3) $-2^{-5} =$

4) $-x^{-7} =$

5) $11x^{-1} =$

6) $-8x^{-3} =$

7) $-12x^{-5} =$

8) $-9x^{-8}y^{-6} =$

9) $32x^{-5}y^{-1} =$

10) $10a^{-9}b^{-3} =$

11) $-17x^4y^{-6} =$

12) $-\dfrac{25}{x^{-5}} =$

13) $-\dfrac{13x}{a^{-7}} =$

14) $\left(-\dfrac{1}{3}\right)^{-4} =$

15) $\left(-\dfrac{3}{4}\right)^{-2} =$

16) $-\dfrac{14}{a^{-6}b^{-3}} =$

17) $-\dfrac{7x}{x^{-8}} =$

18) $-\dfrac{a^{-9}}{b^{-5}} =$

19) $-\dfrac{11}{x^{-5}} =$

20) $\dfrac{8b}{-16c^{-6}} =$

21) $\dfrac{12ab}{a^{-4}b^{-3}} =$

22) $-\dfrac{8n^{-4}}{32p^{-7}} =$

23) $\dfrac{16ab^{-6}}{-6c^{-5}} =$

24) $\left(\dfrac{10a}{5c}\right)^{-4} =$

25) $\left(-\dfrac{12x}{4yz}\right)^{-3} =$

26) $\dfrac{8ab^{-7}}{-5c^{-3}} =$

27) $\left(-\dfrac{x^4}{x^5}\right)^{-5} =$

28) $\left(-\dfrac{x^{-2}}{7x^3}\right)^{-2} =$

29) $\left(-\dfrac{x^{-4}}{x^2}\right)^{-6} =$

WWW.MathNotion.Com

GED Subject Test – Mathematics

Scientific Notation

✎ Write each number in scientific notation.

1) $0.223 =$

2) $0.09 =$

3) $4.5 =$

4) $900 =$

5) $2,000 =$

6) $0.006 =$

7) $33 =$

8) $9,400 =$

9) $1,470 =$

10) $52,000 =$

11) $8,000,000 =$

12) $0.00009 =$

13) $2,158,000 =$

14) $0.0039 =$

15) $0.000075 =$

16) $4,300,000 =$

17) $130,000 =$

18) $4,000,000,000 =$

19) $0.00009 =$

20) $0.0039 =$

✎ Write each number in standard notation.

21) $4 \times 10^{-1} =$

22) $1.2 \times 10^{-3} =$

23) $2.7 \times 10^{5} =$

24) $6 \times 10^{-4} =$

25) $3.6 \times 10^{-3} =$

26) $5.5 \times 10^{5} =$

27) $3.2 \times 10^{4} =$

28) $3.88 \times 10^{6} =$

29) $7 \times 10^{-6} =$

30) $4.2 \times 10^{-7} =$

WWW.MathNotion.Com

GED Subject Test – Mathematics

Square Roots

✎ **Find the value each square root.**

1) $\sqrt{16} =$ _____

2) $\sqrt{25} =$ _____

3) $\sqrt{1} =$ _____

4) $\sqrt{64} =$ _____

5) $\sqrt{0} =$ _____

6) $\sqrt{196} =$ _____

7) $\sqrt{4} =$ _____

8) $\sqrt{256} =$ _____

9) $\sqrt{36} =$ _____

10) $\sqrt{289} =$ _____

11) $\sqrt{169} =$ _____

12) $\sqrt{144} =$ _____

13) $\sqrt{100} =$ _____

14) $\sqrt{1,600} =$ _____

15) $\sqrt{2,500} =$ _____

16) $\sqrt{324} =$ _____

17) $\sqrt{529} =$ _____

18) $\sqrt{20} =$ _____

19) $\sqrt{625} =$ _____

20) $\sqrt{18} =$ _____

21) $\sqrt{50} =$ _____

22) $\sqrt{1,024} =$ _____

23) $\sqrt{160} =$ _____

24) $\sqrt{32} =$ _____

✎ **Evaluate.**

25) $\sqrt{4} \times \sqrt{25} =$ _____

26) $\sqrt{36} \times \sqrt{49} =$ _____

27) $\sqrt{6} \times \sqrt{6} =$ _____

28) $\sqrt{13} \times \sqrt{13} =$ _____

29) $2\sqrt{5} \times 3\sqrt{5} =$ _____

30) $\sqrt{12} \times \sqrt{3} =$ _____

31) $\sqrt{13} + \sqrt{13} =$ _____

32) $\sqrt{10} + 2\sqrt{10} =$ _____

33) $12\sqrt{7} - 10\sqrt{7} =$ _____

34) $4\sqrt{10} \times 2\sqrt{10} =$ _____

35) $5\sqrt{3} \times 8\sqrt{3} =$ _____

36) $6\sqrt{3} - \sqrt{12} =$ _____

GED Subject Test – Mathematics

Simplifying Radical Expressions

✎ **Simplify.**

1) $\sqrt{13x^2} =$

2) $\sqrt{75x^2} =$

3) $\sqrt[3]{27a} =$

4) $\sqrt{64x^5} =$

5) $\sqrt{216a} =$

6) $\sqrt[3]{63w^3} =$

7) $\sqrt{192x} =$

8) $\sqrt{125v} =$

9) $\sqrt[3]{128x^2} =$

10) $\sqrt{100x^9} =$

11) $\sqrt{16x^4} =$

12) $\sqrt[3]{500a^5} =$

13) $\sqrt{242} =$

14) $\sqrt{392p^3} =$

15) $\sqrt{8m^6} =$

16) $\sqrt{198x^3y^3} =$

17) $\sqrt{121x^5y^5} =$

18) $\sqrt{16a^6b^3} =$

19) $\sqrt{90x^5y^7} =$

20) $\sqrt[3]{64y^2x^6} =$

21) $10\sqrt{16x^4} =$

22) $6\sqrt{81x^2} =$

23) $\sqrt[3]{56x^2y^6} =$

24) $\sqrt[3]{1,000x^5y^7} =$

25) $8\sqrt{50a} =$

26) $\sqrt[4]{625x^8y} =$

27) $\sqrt{24x^4y^5r^3} =$

28) $5\sqrt{36x^4y^5z^8} =$

29) $3\sqrt[3]{343x^9y^7} =$

30) $5\sqrt{81a^5b^2c^9} =$

31) $\sqrt[4]{625x^8y^{16}} =$

WWW.MathNotion.Com

GED Subject Test – Mathematics

Answers of Worksheets

Multiplication Property of Exponents

1) 4^6
2) 8^5
3) 7^6
4) 9^4
5) 2^7
6) 5^7
7) 4^7
8) $5x^2$
9) x^6
10) x^9
11) x^9
12) $30x^2$
13) $16x^6$
14) $7x^4$
15) $12x^6$
16) $5x^8$
17) $4x^9$
18) $6x^6$
19) $24x^8$
20) $20x^4y$
21) $7x^{10}y^5$
22) x^7y^7
23) $12x^8y^4$
24) $36x^6y^5$
25) $30x^{11}y^6$
26) $32x^4y^9$
27) $12x^4y^8$
28) $16x^7y^{10}$
29) $21x^3y^{13}$
30) $4x^7y^4$
31) $27x^6y^8$
32) $120x^9y^6$

Zero and Negative Exponents

1) 1
2) $\frac{1}{4}$
3) 0
4) 1
5) $\frac{1}{25}$
6) $\frac{1}{27}$
7) $\frac{1}{9}$
8) $\frac{1}{100}$
9) $\frac{1}{144}$
10) $\frac{1}{32}$
11) $\frac{1}{81}$
12) $\frac{1}{16}$
13) $\frac{1}{216}$
14) $\frac{1}{1,000}$
15) $\frac{1}{30}$
16) $\frac{1}{225}$
17) $\frac{1}{64}$
18) $\frac{1}{128}$
19) $\frac{1}{125}$
20) $\frac{1}{256}$
21) $\frac{1}{243}$
22) $\frac{1}{10,000}$
23) $\frac{1}{1,024}$
24) $\frac{1}{512}$
25) $\frac{1}{400}$
26) $\frac{1}{196}$
27) $\frac{1}{729}$
28) $\frac{1}{10,000}$
29) $\frac{1}{625}$
30) $\frac{1}{4,096}$
31) 64
32) 36
33) 49
34) $\frac{27}{8}$
35) 169
36) $\frac{144}{49}$
37) 216
38) $90,000$
39) $\frac{81}{4}$
40) $\frac{5}{7}$
41) 1
42) $1,024$

Division Property of Exponents

1) $\frac{1}{5}$
2) 8^2
3) 4^4
4) $\frac{1}{3^4}$
5) $\frac{1}{x^5}$
6) $\frac{1}{3^4}$
7) 9^2
8) 10
9) 1
10) $\frac{1}{2x^5}$
11) $\frac{3x^5}{4}$
12) $\frac{3}{2x}$
13) $\frac{7x^5}{y^9}$
14) $\frac{9y^3}{x^4}$

WWW.MathNotion.Com

GED Subject Test – Mathematics

15) $\frac{2x^6}{9}$

16) $\frac{7y^6}{x}$

17) $\frac{2}{x^4 y^{12}}$

18) $\frac{5}{x^2}$

19) $\frac{1}{2x^5 y^3}$

20) $\frac{1}{7}$

21) $\frac{9}{4x^6}$

Powers of Products and Quotients

1) 4^6
2) 2^{12}
3) 2^8
4) 5^{36}
5) 19^{18}
6) 2^{28}
7) 5^6
8) 4^{16}
9) $64x^{10}$
10) $81x^8 y^{16}$
11) $49x^{10} y^4$
12) $125x^{12} y^{12}$
13) $32x^{15} y^{15}$
14) $1{,}000x^9 y^{12}$
15) $169y^8$
16) $25x^{20}$
17) $216x^{21} y^{18}$
18) $144x^{24}$
19) $256x^{20}$
20) $32x^{20} y^{15}$
21) $225x^{14} y^4$
22) $512x^9 y^{15}$
23) $1{,}296x^4 y^8$
24) $\frac{16}{x^8}$
25) x^9
26) $\frac{216 y^3}{x^{12}}$
27) $\frac{1}{x^6 y^{12}}$
28) $x^6 y^6$
29) $\frac{25 y^{16}}{x^4}$
30) $\frac{16}{y^{12}}$

Negative Exponents and Negative Bases

1) $-\frac{1}{9}$
2) $-\frac{1}{81}$
3) $-\frac{1}{32}$
4) $-\frac{1}{x^7}$
5) $\frac{11}{x}$
6) $-\frac{8}{x^3}$
7) $-\frac{12}{x^5}$
8) $-\frac{9}{x^8 y^6}$
9) $\frac{32}{x^5 y}$
10) $\frac{10}{a^9 b^3}$
11) $-\frac{17 x^4}{y^6}$
12) $-25x^5$
13) $-13xa^7$
14) 81
15) $\frac{16}{9}$
16) $-14a^6 b^3$
17) $-7x^9$
18) $-\frac{b^5}{a^9}$
19) $-11x^5$
20) $-\frac{bc^6}{2}$
21) $12a^5 b^4$
22) $-\frac{p^7}{4n^4}$
23) $-\frac{8ac^5}{3b^6}$
24) $\frac{c^4}{16a^4}$
25) $\frac{y^3 z^3}{27 x^3}$
26) $-\frac{8ac^3}{5b^7}$
27) $-x^5$
28) $49x^{10}$
29) x^{36}

Scientific Notation

1) 2.23×10^{-1}
2) 9×10^{-2}
3) 4.5×10^0
4) 9×10^2
5) 2×10^3
6) 6×10^{-3}
7) 3.3×10^1
8) 9.4×10^3
9) 1.47×10^3

WWW.MathNotion.Com

GED Subject Test – Mathematics

10) 5.2×10^4
11) 8×10^6
12) 9×10^{-5}
13) 2.158×10^6
14) 3.9×10^{-3}
15) 7.5×10^{-5}
16) 4.3×10^6

17) 1.3×10^5
18) 4×10^9
19) 9×10^{-5}
20) 3.9×10^{-3}
21) 0.4
22) 0.0012
23) $270,000$

24) 0.0006
25) 0.0036
26) $550,000$
27) $32,000$
28) $3,880,000$
29) 0.000007
30) 0.00000042

Square Roots

1) 4
2) 5
3) 1
4) 8
5) 0
6) 14
7) 2
8) 16
9) 6

10) 17
11) 13
12) 12
13) 10
14) 40
15) 50
16) 18
17) 23
18) $2\sqrt{5}$

19) 25
20) $3\sqrt{2}$
21) $5\sqrt{2}$
22) 32
23) $4\sqrt{10}$
24) $4\sqrt{2}$
25) 10
26) 42
27) 6

28) 13
29) 30
30) 6
31) $2\sqrt{13}$
32) $3\sqrt{10}$
33) $2\sqrt{7}$
34) 80
35) 120
36) $4\sqrt{3}$

Simplifying radical expressions

1) $x\sqrt{13}$
2) $5x\sqrt{3}$
3) $3\sqrt[3]{a}$
4) $8x^2\sqrt{x}$
5) $6\sqrt{6a}$
6) $w\sqrt[3]{63}$
7) $8\sqrt{3x}$
8) $5\sqrt{5v}$
9) $4\sqrt[3]{2x^2}$
10) $10x^4\sqrt{x}$
11) $4x^2$

12) $5a\sqrt[3]{4a^2}$
13) $11\sqrt{2}$
14) $14p\sqrt{2p}$
15) $2m^3\sqrt{2}$
16) $3x.y\sqrt{22xy}$
17) $11x^2y^2\sqrt{xy}$
18) $4a^3b\sqrt{b}$
19) $3x^2y^3\sqrt{10xy}$
20) $4x^2\sqrt[3]{y^2}$
21) $40x^2$
22) $54x$

23) $2y^2\sqrt[3]{7x^2}$
24) $10xy^2\sqrt[3]{x^2y}$
25) $40\sqrt{2a}$
26) $5x^2\sqrt[4]{y}$
27) $2x^2y^2r\sqrt{6yr}$
28) $30x^2y^2z^4\sqrt{y}$
29) $21x^3y^2\sqrt[3]{y}$
30) $45a^2bc^4\sqrt{ac}$
31) $5x^2y^4$

GED Subject Test – Mathematics

Chapter 5:
Algebraic Expressions

Topics that you'll practice in this chapter:

- ✓ Simplifying Variable Expressions
- ✓ Simplifying Polynomial Expressions
- ✓ Translate Phrases into an Algebraic Statement
- ✓ The Distributive Property
- ✓ Evaluating One Variable Expressions
- ✓ Evaluating Two Variables Expressions
- ✓ Combining like Terms

Mathematics is, as it were, a sensuous logic, and relates to philosophy as do the arts, music, and plastic art to poetry. — *K. Shegel*

GED Subject Test – Mathematics

Simplifying Variable Expressions

✎ Simplify each expression.

1) $3(x + 5) =$

2) $(-4)(7x - 5) =$

3) $11x + 5 - 6x =$

4) $-4 - 2x^2 - 6x^2 =$

5) $7 + 13x^2 + 3 =$

6) $3x^2 + 7x + 15x^2 =$

7) $3x^2 - 12x^2 + 4x =$

8) $4x^2 - 8x - 2x =$

9) $6x + 7(3 - 4x) =$

10) $8x + 4(15x - 3) =$

11) $6(-3x - 9) - 17 =$

12) $-11x^2 - (-5x) =$

13) $2x + 7 + 5 - 8x =$

14) $7 + 6x - 11 - 5x =$

15) $27x + 8 - 13 - 5x =$

16) $(-11)(-5x + 2) - 41x =$

17) $19x - 4(4 - 2x) =$

18) $16x + 3(3x + 6) + 10 =$

19) $5(-2x - 4) - 13x =$

20) $16x - 3x(x + 10) =$

21) $17x + 5x(2 - 4x) =$

22) $5x(-4x - 7) + 20x =$

23) $25x - 19 + 4x^2 =$

24) $6x(x - 11) + 25 =$

25) $4x - 5 + 15x + 3x^2 =$

26) $-7x^2 - 11x - 9x =$

27) $10x - 9x^2 - 3x^2 - 7 =$

28) $13 + 3x^2 - 9x^2 - 21x =$

29) $22x + 10x^2 - 15x + 17 =$

30) $4x^2 + 25x + 21x^2 =$

31) $29 - 12x^2 - 23x - 4x^2 =$

32) $22x - 19x - 9x^2 + 30 =$

WWW.MathNotion.Com

GED Subject Test – Mathematics

Simplifying Polynomial Expressions

✎ **Simplify each polynomial.**

1) $(2x^3 + 8x^2) - (11x + 3x^2) =$ _____

2) $(2x^5 + 7x^3) - (5x^3 + 11x^2) =$ _____

3) $(41x^4 + 5x^2) - (4x^2 + 20x^4) =$ _____

4) $13x - 8x^2 + 4(4x^2 + 3x^3) =$ _____

5) $(4x^3 - 22) + 5(3x^2 - 6x^3) =$ _____

6) $(4x^3 - 3x) - 5(2x^3 + x^4) =$ _____

7) $5(5x - 2x^3) - 2(8x^3 + 5x^2) =$ _____

8) $(3x^2 - 10x) - (5x^3 + 14x^2) =$ _____

9) $5x^3 - (3x^4 + 5x) + 2x^2 =$ _____

10) $11x^4 - (3x^2 + 5x) + 7x =$ _____

11) $(6x^2 - 3x^4) - (10x^4 + 3x^2) =$ _____

12) $2x^2 - 7x^3 + 19x^4 - 22x^3 =$ _____

13) $10x^2 - x^4 + 4x^4 - 32x^3 =$ _____

14) $-5x^2 + 17x^3 - 8x^2 - 6x =$ _____

15) $x^4 - 11x^5 - 30x^4 + 5x^2 =$ _____

16) $21x^3 + 13x - 5x^2 - 11x^3 =$ _____

WWW.MathNotion.Com

GED Subject Test – Mathematics

Translate Phrases into an Algebraic Statement

✎ **Write an algebraic expression for each phrase.**

1) 9 multiplied by x. _____

2) Subtract 11 from y. _____

3) 19 divided by x. _____

4) 38 decreased by y. _____

5) Add y to 40. _____

6) The square of 6. _____

7) x raised to the fifth power. _____

8) The sum of six and a number. _____

9) The difference between fifty-seven and y. _____

10) The quotient of nine and a number. _____

11) The quotient of the square of x and 25. _____

12) The difference between x and 6 is 19. _____

13) 10 times a reduced by the square of b. _____

14) Subtract the product of a and b from 41. _____

GED Subject Test – Mathematics

The Distributive Property

✎ Use the distributive property to simply each expression.

1) $4(1 + 2x) =$

2) $2(4 + 7x) =$

3) $3(4x - 4) =$

4) $(2x - 5)(-6) =$

5) $(-3)(x + 6) =$

6) $(4 + 3x)2 =$

7) $(-5)(8 - 3x) =$

8) $-(-5 - 7x) =$

9) $(-6x + 3)(-3) =$

10) $(-4)(x - 7) =$

11) $-(5 - 3x) =$

12) $3(9 + 4x) =$

13) $6(4 + 3x) =$

14) $(-5x + 3)2 =$

15) $(5 - 8x)(-3) =$

16) $(-12)(3x + 3) =$

17) $(5 - 3x)6 =$

18) $4(2 + 6x) =$

19) $8(7x - 3) =$

20) $(-2x + 3)4 =$

21) $(7 - 5x)(-9) =$

22) $(-10)(x - 8) =$

23) $(11 - 4x)3 =$

24) $(-6)(10x - 4) =$

25) $(3 - 9x)(-7) =$

26) $(-9)(x + 9) =$

27) $(-3 + 5x)(-7) =$

28) $(-5)(8 - 10x) =$

29) $12(4x - 8) =$

30) $(-10x + 13)(-3) =$

31) $(-8)(3x - 2) + 4(x + 5) =$

32) $(-8)(x + 4) - (6 + 5x) =$

WWW.MathNotion.Com

GED Subject Test – Mathematics

Evaluating One Variable Expressions

✏️ **Evaluate each expression using the value given.**

1) $8 - x$, $x = 5$

2) $x - 9$, $x = 5$

3) $5x + 4$, $x = 3$

4) $x - 13$, $x = -4$

5) $12 - x$, $x = 4$

6) $x + 2$, $x = 6$

7) $4x + 8$, $x = 3$

8) $x + (-7)$, $x = -8$

9) $4x + 5$, $x = 2$

10) $3x + 9$, $x = -2$

11) $15 + 3x - 7$, $x = 2$

12) $17 - 3x$, $x = 3$

13) $8x - 9$, $x = 4$

14) $5x + 4$, $x = -3$

15) $10x + 5$, $x = 3$

16) $14 - 4x$, $x = -6$

17) $3(5x + 3)$, $x = 9$

18) $4(-3x - 6)$, $x = 3$

19) $7x - 2x + 12$, $x = 4$

20) $(5x + 6) \div 2$, $x = 8$

21) $(x + 18) \div 10$, $x = 12$

22) $5x - 12 + 3x$, $x = -3$

23) $(6 - 4x)(-3)$, $x = -4$

24) $9x^2 + 3x - 6$, $x = 2$

25) $x^2 - 10x$, $x = -5$

26) $3x(7 - 2x)$, $x = 2$

27) $12x + 6 - 2x^2$, $x = -4$

28) $(-3)(4x - 8 + 3x)$, $x = 3$

29) $(-6) + \frac{x}{4} + 3x$, $x = 16$

30) $(-6) + \frac{x}{5}$, $x = 35$

31) $\left(-\frac{45}{x}\right) - 7 + 2x$, $x = 9$

32) $\left(-\frac{21}{x}\right) - 12 + 4x$, $x = 7$

GED Subject Test – Mathematics

Evaluating Two Variables Expressions

✎ **Evaluate each expression using the values given.**

1) $2x - 4y$,

   $x = 4, y = 1$

2) $3x + 5y$,

   $x = -2, y = 2$

3) $-7a + 4b$,

   $a = 2, b = 4$

4) $3x + 5 - y$,

   $x = 5, y = 6$

5) $3z + 12 - 2k$,

   $z = 5, k = 6$

6) $6(-x - 3y)$,

   $x = 5, y = -2$

7) $5a + 3b$,

   $a = 3, b = 4$

8) $7x \div 3y$,

   $x = 3, y = 7$

9) $2x + 15 + 5y$,

   $x = -3, y = 1$

10) $5a - (18 - b)$,

    $a = 2, b = 8$

11) $2z + 20 + 5k$,

    $z = -6, k = 5$

12) $xy + 10 + 4x$,

    $x = 3, y = 5$

13) $2x + 4y - 8 + 5$,

    $x = 5, y = 2$

14) $\left(-\frac{24}{x}\right) + 3 + 2y$,

    $x = 4, y = 6$

15) $(-3)(-3a - 3b)$,

    $a = 4, b = 5$

16) $12 + 4x - 7 - y$,

    $x = 3, y = 5$

17) $11x + 5 - 8y + 6$,

    $x = 5, y = 2$

18) $10 + 2(-4x - 5y)$,

    $x = 5, y = 4$

19) $5x + 13 + 6y$,

    $x = 5, y = 6$

20) $10a - (7a + 3b) - 11$,

    $a = 3, b = 8$

WWW.MathNotion.Com

GED Subject Test – Mathematics

Combining like Terms

✎ **Simplify each expression.**

1) $11x + 3x + 6 =$

2) $8(2x - 6) =$

3) $18x - 7x + 11 =$

4) $(-4)(6x - 7) =$

5) $22x - 10x - 5 =$

6) $32x - 13 + 8x =$

7) $15 - (8x - 11) =$

8) $-24x + 17 - 11x =$

9) $12x - 8 - 6x + 9 =$

10) $21x + 5 - 36 + 12x =$

11) $28x + 3x - 11 =$

12) $(-3x + 4)5 =$

13) $2 + 4x + 9x - 8 =$

14) $6(2x - 5x) - 4 =$

15) $4(5x + 11) + 3x =$

16) $x - 14 - 11x =$

17) $5(10 + 9x) - 8x =$

18) $42x + 17 - 23x =$

19) $(-7x) + 19 + 20x =$

20) $(-7x) - 33 + 29x =$

21) $4(5x + 3) - 19x =$

22) $5(6 - 2x) - 15x =$

23) $-24x + (11 - 18x) =$

24) $(-9) - (6)(7x + 3) =$

25) $(-1)(8x - 10) - 21x =$

26) $-36x + 14 + 27x - 5x =$

27) $3(-13x + 6) - 17x =$

28) $-5x - 42 + 32x =$

29) $37x - 19x + 15 - 9x =$

30) $3(5x + 7x) - 31 =$

31) $14 - 6x - 15 - 9x =$

32) $-2(-5x - 7x) + 27x =$

WWW.MathNotion.Com

GED Subject Test – Mathematics

Answers of Worksheets

Simplifying Variable Expressions

1) $3x + 15$
2) $-28x + 20$
3) $5x + 5$
4) $-8x^2 - 4$
5) $13x^2 + 10$
6) $18x^2 + 7x$
7) $-9x^2 + 4x$
8) $4x^2 - 10x$
9) $-22x + 21$
10) $68x - 12$
11) $-18x - 71$
12) $-11x^2 + 5x$
13) $-6x + 12$
14) $x - 4$
15) $22x - 5$
16) $14x - 22$
17) $27x - 16$
18) $25x + 28$
19) $-23x - 20$
20) $-3x^2 - 14x$
21) $-20x^2 + 27x$
22) $-20x^2 - 15x$
23) $4x^2 + 25x - 19$
24) $6x^2 - 66x + 25$
25) $3x^2 + 19x - 5$
26) $-7x^2 - 20x$
27) $-12x^2 + 10x - 7$
28) $-6x^2 - 21x + 13$
29) $10x^2 + 7x + 17$
30) $25x^2 + 25x$
31) $-16x^2 - 23x + 29$
32) $-9x^2 + 3x + 30$

Simplifying Polynomial Expressions

1) $2x^3 + 5x^2 - 11x$
2) $2x^5 + 2x^3 - 11x^2$
3) $21x^4 + x^2$
4) $12x^3 + 8x^2 + 13x$
5) $-26x^3 + 15x^2 - 22$
6) $-5x^4 - 6x^3 - 3x$
7) $-26x^3 - 10x^2 + 25x$
8) $-5x^3 - 11x^2 - 10x$
9) $-3x^4 + 5x^3 + 2x^2 - 5x$
10) $11x^4 - 3x^2 + 2x$
11) $-13x^4 + 3x^2$
12) $19x^4 - 29x^3 + 2x^2$
13) $3x^4 - 32x^3 + 10x^2$
14) $17x^3 - 13x^2 - 6x$
15) $-11x^5 - 29x^4 + 5x^2$
16) $10x^3 - 5x^2 + 13x$

Translate Phrases into an Algebraic Statement

1) $9x$
2) $y - 11$
3) $\frac{19}{x}$
4) $38 - y$
5) $y + 40$
6) 6^2
7) x^5
8) $6 + x$
9) $57 - y$
10) $\frac{9}{x}$
11) $\frac{x^2}{25}$
12) $x - 6 = 19$
13) $10a - b^2$
14) $41 - ab$

The Distributive Property

1) $8x + 4$
2) $14x + 8$
3) $12x - 12$
4) $-12x + 30$
5) $-3x - 18$
6) $6x + 8$
7) $15x - 40$
8) $7x + 5$
9) $18x - 9$
10) $-4x + 28$
11) $3x - 5$
12) $12x + 27$

GED Subject Test – Mathematics

13) $18x + 24$	18) $24x + 8$	23) $-12x + 33$	28) $50x - 40$
14) $-10x + 6$	19) $56x - 24$	24) $-60x + 24$	29) $48x - 96$
15) $24x - 15$	20) $-8x + 12$	25) $63x - 21$	30) $30x - 39$
16) $-36x - 36$	21) $45x - 63$	26) $-9x - 81$	31) $-20x + 36$
17) $-18x + 30$	22) $-10x + 80$	27) $-35x + 21$	32) $-13x - 38$

Evaluating One Variables

1) 3	9) 13	17) 144	25) 75
2) -4	10) 3	18) -60	26) 18
3) 19	11) 14	19) 32	27) -74
4) -17	12) 8	20) 23	28) -39
5) 8	13) 23	21) 3	29) 46
6) 8	14) -11	22) -36	30) 1
7) 20	15) 35	23) -66	31) 6
8) -15	16) 38	24) 36	32) 13

Evaluating Two Variables

1) 4	6) 6	11) 33	16) 12
2) 4	7) 27	12) 37	17) 50
3) 2	8) 1	13) 15	18) -70
4) 14	9) 14	14) 9	19) 74
5) 15	10) 0	15) 81	20) -26

Combining like Terms

1) $14x + 6$	9) $6x + 1$	17) $37x + 50$	25) $-29x + 10$
2) $16x - 48$	10) $33x - 31$	18) $19x + 17$	26) $-14x + 14$
3) $11x + 11$	11) $31x - 11$	19) $13x + 19$	27) $-56x + 18$
4) $-24x + 28$	12) $-15x + 20$	20) $22x - 33$	28) $27x - 42$
5) $12x - 5$	13) $13x - 6$	21) $x + 12$	29) $9x + 15$
6) $40x - 13$	14) $-18x - 4$	22) $-25x + 30$	30) $36x - 31$
7) $-8x + 26$	15) $23x + 44$	23) $-42x + 11$	31) $-15x - 1$
8) $-35x + 17$	16) $-10x - 14$	24) $-42x - 27$	32) $51x$

WWW.MathNotion.Com

GED Subject Test – Mathematics

Chapter 6 :
Equations and Inequalities

Topics that you'll practice in this chapter:

- ✓ One–Step Equations
- ✓ Multi–Step Equations
- ✓ Graphing Single–Variable Inequalities
- ✓ One–Step Inequalities
- ✓ Multi-Step Inequalities
- ✓ Systems of Equations
- ✓ Systems of Equations Word Problems

"Life is a math equation. In order to gain the most, you have to know how to convert negatives into positives." – Anonymous

GED Subject Test – Mathematics

One–Step Equations

✎ **Find the answer for each equation.**

1) $3x = 90, x = $ ____

2) $5x = 35, x = $ ____

3) $6x = 24, x = $ ____

4) $24x = 144, x = $ ____

5) $x + 15 = 20, x = $ ____

6) $x - 7 = 4, x = $ ____

7) $x - 9 = 2, x = $ ____

8) $x + 15 = 23, x = $ ____

9) $x - 4 = 13, x = $ ____

10) $12 = 16 + x, x = $ ____

11) $x - 10 = 2, x = $ ____

12) $5 - x = -11, x = $ ____

13) $28 = -6 + x, x = $ ____

14) $x - 20 = -35, x = $ ____

15) $x + 14 = -4, x = $ ____

16) $14 = 28 - x, x = $ ____

17) $7 + x = -7, x = $ ____

18) $x - 16 = 4, x = $ ____

19) $30 = x - 15, x = $ ____

20) $x - 5 = -18, x = $ ____

21) $x - 10 = 24, x = $ ____

22) $x - 20 = -25, x = $ ____

23) $x - 17 = 30, x = $ ____

24) $-70 = x - 28, x = $ ____

25) $x - 9 = 13, x = $ ____

26) $36 = 4x, x = $ ____

27) $x - 35 = 25, x = $ ____

28) $x - 25 = 10, x = $ ____

29) $70 - x = 16, x = $ ____

30) $x - 10 = 14, x = $ ____

31) $17 - x = -13, x = $ ____

32) $x - 9 = -30, x = $ ____

WWW.MathNotion.Com

GED Subject Test – Mathematics

Multi–Step Equations

✎ Find the answer for each equation.

1) $3x + 3 = 9$

2) $-x + 5 = 12$

3) $4x - 8 = 8$

4) $-(3 - x) = 5$

5) $4x - 8 = 16$

6) $12x - 15 = 9$

7) $2x - 18 = 2$

8) $4x + 8 = 16$

9) $24x + 27 = 75$

10) $-14(3 + x) = 14$

11) $-3(2 + x) = 6$

12) $12 = -(x - 7)$

13) $3(3 - x) = 30$

14) $-15 = -(3x + 6)$

15) $40(3 + x) = 40$

16) $5(x - 10) = 25$

17) $-18 = x + 8x$

18) $3x + 25 = -2x - 10$

19) $7(6 + 3x) = -63$

20) $18 - 3x = -4 - 5x$

21) $4 - 6x = 36 + 2x$

22) $15 + 15x = -5 + 5x$

23) $42 = (-6x) - 7 + 7$

24) $21 = 3x - 21 + 4x$

25) $-18 = -6x - 9 + 3x$

26) $5x - 15 = -29 + 6x$

27) $7x - 18 = 4x + 3$

28) $-7 - 4x = 5(4 - x)$

29) $x - 5 = -5(-3 - x)$

30) $13x - 68 = 15x - 102$

31) $-5x - 3 = -3(9 + 3x)$

32) $-2x - 15 = 6x + 17$

WWW.MathNotion.Com

GED Subject Test – Mathematics

Graphing Single–Variable Inequalities

✎ **Draw a graph for each inequality.**

1) $x > -1$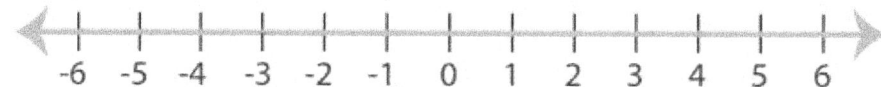

2) $x \leq 2$

3) $x \geq 0$

4) $x < -3$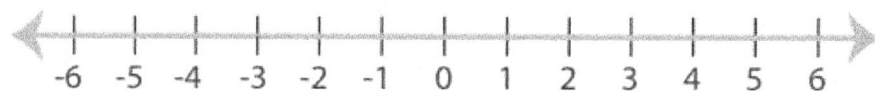

5) $x < \frac{1}{2}$

6) $x \leq -2$

7) $x \leq 3$

8) $x \geq -\frac{7}{2}$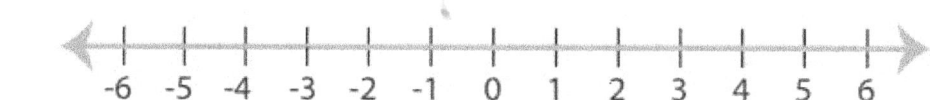

GED Subject Test – Mathematics

One–Step Inequalities

✎ **Find the answer for each inequality and graph it.**

1) $x + 4 \geq 4$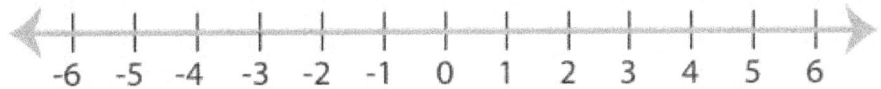

2) $x - 5 \leq 2$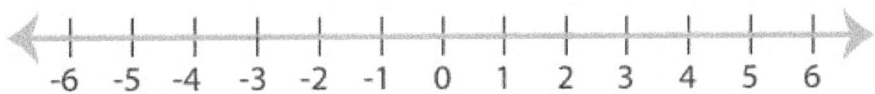

3) $5x > 35$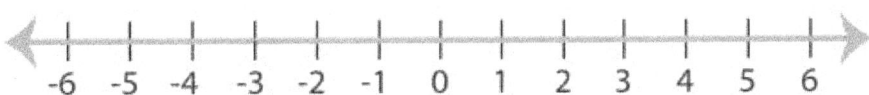

4) $9 + x \leq 11$

5) $x - 5 < -9$

6) $9x \geq 72$

7) $9x \leq 27$

8) $x + 19 > 16$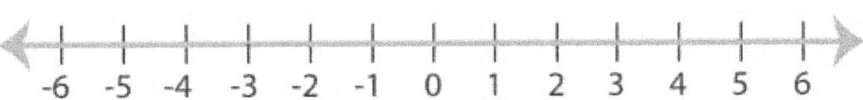

GED Subject Test – Mathematics

Multi-Step Inequalities

✎ **Calculate each inequality.**

1) $x - 3 \leq 7$

2) $8 - x \leq 8$

3) $3x - 9 \leq 9$

4) $4x - 4 \geq 8$

5) $x - 7 \geq 1$

6) $5x - 15 \leq 5$

7) $6x - 8 \leq 4$

8) $-11 + 6x \leq 12$

9) $4(x - 4) \leq 16$

10) $3x - 10 \leq 11$

11) $5x - 25 < 25$

12) $9x - 5 < 22$

13) $20 - 7x \geq -15$

14) $33 + 6x < 45$

15) $8 + 8x \geq 96$

16) $7 + 3x < 13$

17) $4x - 3 < 9$

18) $5(2 - 2x) \geq -30$

19) $-(7 + 6x) < 29$

20) $12 - 8x \geq -20$

21) $-4(x - 6) > 24$

22) $\dfrac{3x + 9}{6} \leq 10$

23) $\dfrac{4x - 10}{3} \leq 2$

24) $\dfrac{2x - 8}{3} > 2$

25) $8 + \dfrac{x}{6} < 9$

26) $\dfrac{9x}{7} - 4 < 5$

27) $\dfrac{15x + 45}{15} > 1$

28) $16 + \dfrac{x}{4} < 6$

WWW.MathNotion.Com

GED Subject Test – Mathematics

Systems of Equations

✎ **Calculate each system of equations.**

1) $-x + y = 2$ $x =$ ___
   $-4x + 2y = 6$ $y =$ ___

2) $-15x + 3y = -9$ $x =$ ___
   $9x - 16y = 48$ $y =$ ___

3) $y = -7$ $x =$ ___
   $6x + 5y = 7$ $y =$ ___

4) $3y = -9x + 15$ $x =$ ___
   $5x - 4y = -3$ $y =$ ___

5) $10x - 9y = -13$ $x =$ ___
   $-5x + 3y = 11$ $y =$ ___

6) $-12x - 16y = 20$ $x =$ ___
   $6x - 12y = 30$ $y =$ ___

7) $5x - 14y = -23$ $x =$ ___
   $-18x + 21y = 24$ $y =$ ___

8) $15x - 21y = -6$ $x =$ ___
   $2x - 3y = -2$ $y =$ ___

9) $-x + 3y = 3$ $x =$ ___
   $-14x + 16y = -10$ $y =$ ___

10) $x + 5y = 50$ $x =$ ___
    $3x + 10y = 80$ $y =$ ___

11) $6x - 7y = -8$ $x =$ ___
    $-x - 4y = -9$ $y =$ ___

12) $2x + 4y = -10$ $x =$ ___
    $2x - 8y = 14$ $y =$ ___

13) $4x + 3y = 12$ $x =$ ___
    $5x - 3y = 15$ $y =$ ___

14) $3x - 2y = 3$ $x =$ ___
    $7x - 8y = 22$ $y =$ ___

15) $3x + 2y = 5$ $x =$ ___
    $-10x - 4y = -14$ $y =$ ___

16) $10x + 7y = 1$ $x =$ ___
    $-5x - 7y = 24$ $y =$ ___

GED Subject Test – Mathematics

Systems of Equations Word Problems

✎ **Find the answer for each word problem.**

1) Tickets to a movie cost $4 for adults and $3 for students. A group of friends purchased 8 tickets for $31.00. How many adults ticket did they buy? _____

2) At a store, Eva bought two shirts and five hats for $77.00. Nicole bought three same shirts and four same hats for $84.00. What is the price of each shirt? _____

3) A farmhouse shelters 18 animals, some are pigs, and some are ducks. Altogether there are 66 legs. How many pigs are there? _____

4) A class of 214 students went on a field trip. They took 36 vehicles, some cars and some buses. If each car holds 5 students and each bus hold 22 students, how many buses did they take? _____

5) A theater is selling tickets for a performance. Mr. Smith purchased 5 senior tickets and 3 child tickets for $105 for his friends and family. Mr. Jackson purchased 3 senior tickets and 5 child tickets for $79. What is the price of a senior ticket? $_____

6) The difference of two numbers is 10. Their sum is 20. What is the bigger number? $_____

7) The sum of the digits of a certain two-digit number is 7. Reversing its digits increase the number by 9. What is the number? _____

8) The difference of two numbers is 11. Their sum is 25. What are the numbers? _____

9) The length of a rectangle is 5 meters greater than 2 times the width. The perimeter of rectangle is 28 meters. What is the length of the rectangle? _____

10) Jim has 25 nickels and dimes totaling $1.80. How many nickels does he have? _____

WWW.MathNotion.Com

GED Subject Test – Mathematics

Answers of Worksheets

One–Step Equations

1) 30	9) 17	17) −14	25) 22
2) 7	10) −4	18) 20	26) 9
3) 4	11) 12	19) 45	27) 60
4) 6	12) 16	20) −13	28) 35
5) 5	13) 34	21) 34	29) 54
6) 11	14) −15	22) −5	30) 24
7) 11	15) −18	23) 47	31) 30
8) 8	16) 14	24) −42	32) −21

Multi–Step Equations

1) 2	9) 2	17) −2	25) 3
2) −7	10) −4	18) −7	26) 14
3) 4	11) −4	19) −5	27) 7
4) 8	12) −5	20) −11	28) 27
5) 6	13) −7	21) −4	29) −5
6) 2	14) 3	22) −2	30) 17
7) 10	15) −2	23) −7	31) −6
8) 2	16) 15	24) 6	32) −4

Graphing Single–Variable Inequalities

1) [number line with open circle at −1, shaded to the right]

2) [number line with closed circle at 2, shaded to the right]

3) [number line with open circle at 0, shaded to the right]

4) [number line with open circle at −3, shaded to the right]

WWW.MathNotion.Com

GED Subject Test – Mathematics

5) number line with open circle at 1/2

6) number line with closed point at -2

7) number line with closed point at 3

8) number line with open circle at -3.5

One–Step Inequalities

1) number line with point at 0
2) number line with point at 7
3) number line with point at 7
4) number line with point at 2
5) number line with point at -4
6) number line with point at 8
7) number line with point at 3
8) number line with open circle at -3

Multi-Step Inequalities

1) $x \leq 10$	7) $x \leq 2$	12) $x < 3$	18) $x \leq 4$
2) $x \geq 0$	8) $x \leq \frac{23}{6}$	13) $x \leq 5$	19) $x > -6$
3) $x \leq 6$	9) $x \leq 8$	14) $x < 2$	20) $x \leq 4$
4) $x \geq 3$	10) $x \leq 7$	15) $x \geq 11$	21) $x < 0$
5) $x \geq 8$	11) $x < 10$	16) $x < 2$	22) $x \leq 17$
6) $x \leq 4$		17) $x < 3$	23) $x \leq 4$

WWW.MathNotion.Com

GED Subject Test – Mathematics

24) $x > 7$ 26) $x < 7$ 28) $x < -40$

25) $x < 6$ 27) $x > -2$

Systems of Equations

1) $x = -1, y = 1$
2) $x = 0, y = -3$
3) $x = 7$
4) $x = 1, y = 2$
5) $x = -4, y = -3$
6) $x = 1, y = -2$
7) $x = 1, y = 2$
8) $x = 8, y = 6$
9) $x = 3, y = 2$
10) $x = -20, y = 14$
11) $x = 1, y = 2$
12) $x = -1, y = -2$
13) $x = 3, y = 0$
14) $x = -2, y = -\frac{9}{2}$
15) $x = 1, y = 1$
16) $x = 5, y = -7$

Systems of Equations Word Problems

1) 7
2) $16
3) 15
4) 2
5) $18
6) 15
7) 34
8) 18, 7
9) 11 meters
10) 14

GED Subject Test – Mathematics

GED Subject Test – Mathematics

Chapter 7 :
Linear Functions

Topics that you'll practice in this chapter:

- ✓ Finding Slope
- ✓ Graphing Lines Using Line Equation
- ✓ Writing Linear Equations
- ✓ Graphing Linear Inequalities
- ✓ Finding Midpoint
- ✓ Finding Distance of Two Points

"Nature is written in mathematical language." – *Galileo Galilei*

GED Subject Test – Mathematics

Finding Slope

✎ **Find the slope of each line.**

1) $y = x + 8$

2) $y = -3x + 5$

3) $y = 2x + 12$

4) $y = -4x + 19$

5) $y = 11 + 6x$

6) $y = 7 - 5x$

7) $y = 8x + 19$

8) $y = -9x + 20$

9) $y = -7x + 4$

10) $y = 3x - 8$

11) $y = \frac{1}{3}x + 8$

12) $y = -\frac{4}{5}x + 9$

13) $-3x + 6y = 30$

14) $4x + 4y = 16$

15) $3y - x = 10$

16) $8y - x = 5$

✎ **Find the slope of the line through each pair of points.**

17) $(2, 3), (7, 10)$

18) $(-3, 5), (2, 15)$

19) $(5, -3), (1, 9)$

20) $(-5, -5), (10, 25)$

21) $(22, 3), (7, 18)$

22) $(-16, 8), (-7, 26)$

23) $(25, 11), (29, 19)$

24) $(26, -19), (14, 17)$

25) $(22, -13), (20, -11)$

26) $(19, 7), (15, -3)$

27) $(5, 7), (11, 19)$

28) $(52, -62), (40, 70)$

GED Subject Test – Mathematics

Graphing Lines Using Line Equation

✏️ Sketch the graph of each line.

1) $y = x - 2$ 2) $y = -3x + 2$ 3) $x + y = 0$

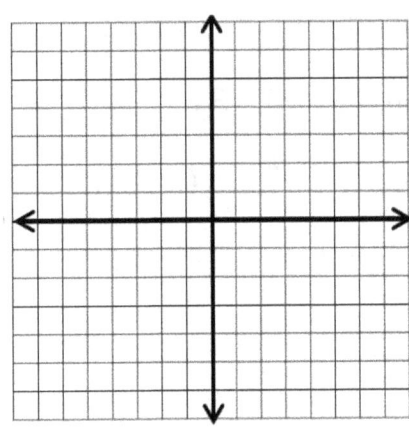 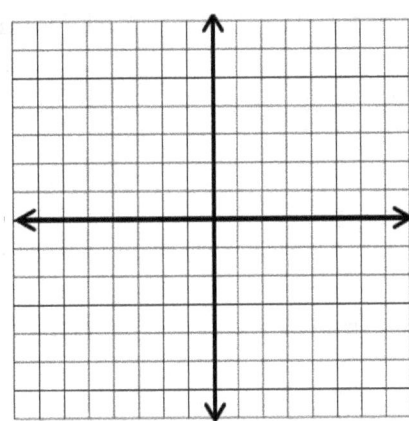 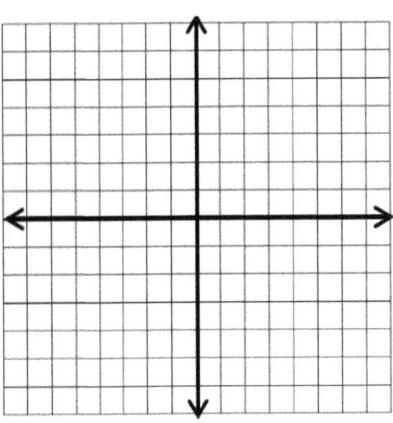

4) $x + y = -3$ 5) $2x + 3y = -4$ 6) $y - 3x + 6 = 0$

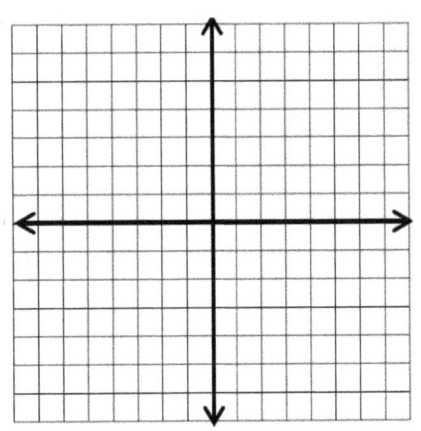 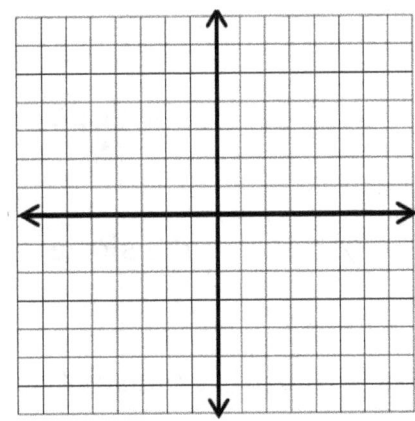 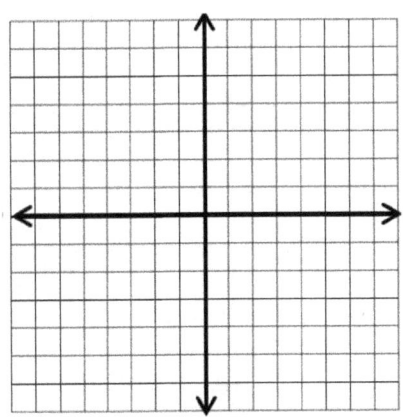

GED Subject Test – Mathematics

Writing Linear Equations

✍ **Write the equation of the line through the given points.**

1) Through: $(2, -5), (3, 9)$

2) Through: $(-6, 3), (3, 12)$

3) Through: $(10, 7), (5, 27)$

4) Through: $(15, 11), (3, -1)$

5) Through: $(24, 17), (12, -7)$

6) Through: $(8, 29), (4, -7)$

7) Through: $(20, -16), (12, 0)$

8) Through: $(-3, 10), (2, -5)$

9) Through: $(-6, 17), (4, -3)$

10) Through: $(-8, 22), (5, -4)$

11) Through: $(9, 27), (3, -3)$

12) Through: $(11, 32), (9, 4)$

13) Through: $(-3, 13), (-4, 0)$

14) Through: $(-5, 5), (5, 15)$

15) Through: $(18, -32), (11, 3)$

16) Through: $(-4, 25), (4, -15)$

✍ **Find the answer for each problem.**

17) What is the equation of a line with slope 6 and intercept 12? _____

18) What is the equation of a line with slope -11 and intercept -4? _____

19) What is the equation of a line with slope -3 and passes through point $(5, 2)$? _____

20) What is the equation of a line with slope -5 and passes through point $(-2, -1)$? _____

21) The slope of a line is -10 and it passes through point $(-3, 0)$. What is the equation of the line? _____

22) The slope of a line is 8 and it passes through point $(0, 7)$. What is the equation of the line? _____

Graphing Linear Inequalities

✏️ **Sketch the graph of each linear inequality.**

1) $y > 4x - 5$ 2) $y < 2x + 4$ 3) $y \leq -5x - 2$

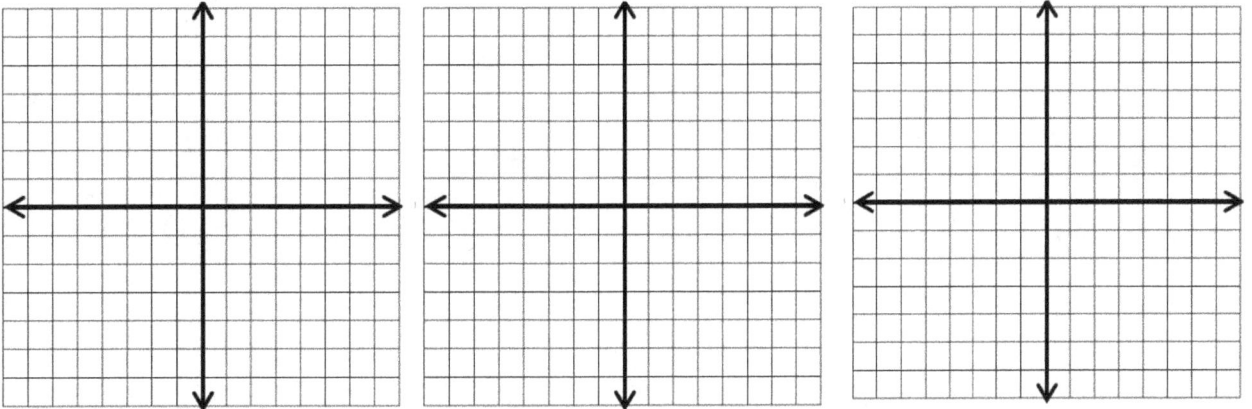

4) $4y \geq 12 + 4x$ 5) $-12y < 3x - 24$ 6) $5y \geq -15x + 10$

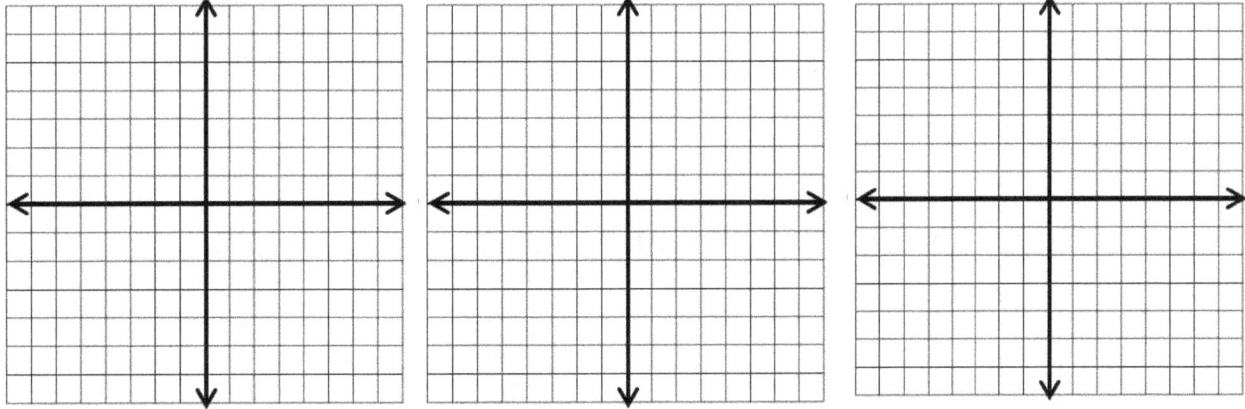

GED Subject Test – Mathematics

Finding Midpoint

✎ **Find the midpoint of the line segment with the given endpoints.**

1) $(-4, -3), (2, 3)$
2) $(9, 0), (-1, 8)$
3) $(9, -6), (3, 14)$
4) $(-10, -6), (0, 8)$
5) $(2, -5), (14, -15)$
6) $(-10, -3), (4, -13)$
7) $(8, 7), (-8, 13)$
8) $(-3, 6), (-9, 2)$
9) $(-4, 5), (16, -9)$
10) $(7, 14), (9, -2)$
11) $(-8, 6), (6, 6)$
12) $(10, 5), (-2, -3)$

13) $(-5, 12), (-3, 3)$
14) $(12, 7), (8, -2)$
15) $(10, 2), (-6, 14)$
16) $(-1, -2), (-7, 10)$
17) $(7, -7), (13, -13)$
18) $(-3, -8), (11, -4)$
19) $(5, -11), (-8, 9)$
20) $(14, -4), (16, 14)$
21) $(0, -5), (8, -1)$
22) $(3, 0), (-21, 18)$
23) $(17, -3), (-7, -5)$
24) $(26, -12), (6, 24)$

✎ **Find the answer for each problem.**

25) One endpoint of a line segment is $(-3, 7)$ and the midpoint of the line segment is $(-6, 9)$. What is the other endpoint? _____

26) One endpoint of a line segment is $(-3, 7)$ and the midpoint of the line segment is $(1, 5)$. What is the other endpoint? _____

27) One endpoint of a line segment is $(-10, -16)$ and the midpoint of the line segment is $(2, 9)$. What is the other endpoint? _____

WWW.MathNotion.Com

GED Subject Test – Mathematics

Finding Distance of Two Points

✎ **Find the distance between each pair of points.**

1) $(6, 3), (-3, -9)$

2) $(5, 2), (-10, -6)$

3) $(8, 5), (8, 3)$

4) $(-8, -2), (2, 22)$

5) $(6, -7), (-3, -7)$

6) $(12, 0), (-9, -20)$

7) $(3, 20), (3, -5)$

8) $(10, 17), (5, 5)$

9) $(7, -2), (-4, -2)$

10) $(13, 4), (5, -2)$

11) $(11, 13), (5, 5)$

12) $(1, 4), (-23, -3)$

13) $(9, 8), (5, -4)$

14) $(-11, -4), (5, 8)$

15) $(-2, -6), (-2, -12)$

16) $(-1, -4), (23, 3)$

17) $(19, 3), (7, -6)$

18) $(-5, -2), (3, 4)$

19) $(2, 6), (2, -12)$

20) $(-4, -2), (8, -2)$

✎ **Find the answer for each problem.**

21) Triangle ABC is a right triangle on the coordinate system and its vertices are $(-2, 5)$, $(-2, 1)$, and $(1, 1)$. What is the area of triangle ABC? _____

22) Three vertices of a triangle on a coordinate system are $(3, -6)$, $(-5, -12)$, and $(3, -18)$. What is the perimeter of the triangle? _____

23) Four vertices of a rectangle on a coordinate system are $(-2, 2)$, $(-2, 6)$, $(4, 2)$, and $(4, 6)$. What is its perimeter? _____

WWW.MathNotion.Com

GED Subject Test – Mathematics

Answers of Worksheets

Finding Slope

1) 1
2) −3
3) 2
4) −4
5) 6
6) −5
7) 8
8) −9
9) −7
10) 3
11) $\frac{1}{3}$
12) $-\frac{4}{5}$
13) $\frac{1}{2}$
14) −1
15) $\frac{1}{3}$
16) $\frac{1}{8}$
17) $\frac{7}{5}$
18) 2
19) −3
20) 2
21) −1
22) 2
23) 2
24) −3
25) −1
26) $\frac{5}{2}$
27) 2
28) −11

Graphing Lines Using Line Equation

1) $y = x - 2$

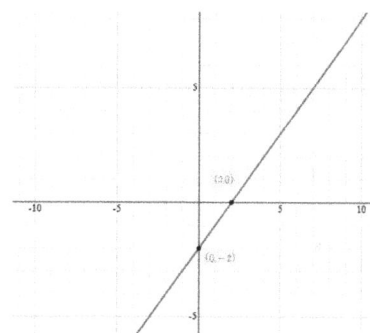

2) $y = -3x + 2$

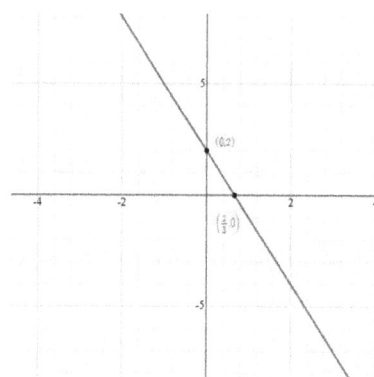

3) $x + y = 0$

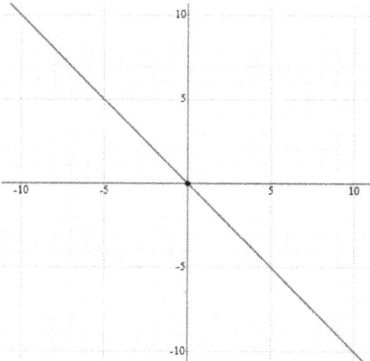

4) $x + y = -3$

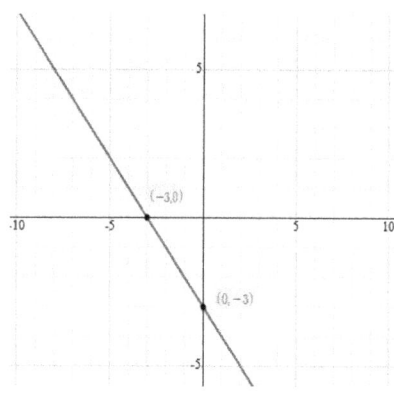

5) $2x + 3y = -4$

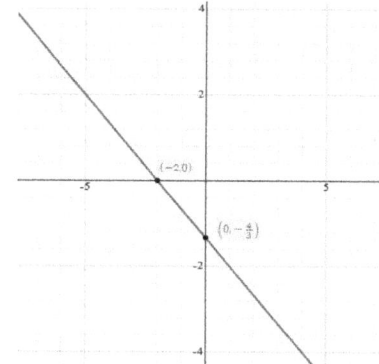

6) $y - 3x + 6 = 0$

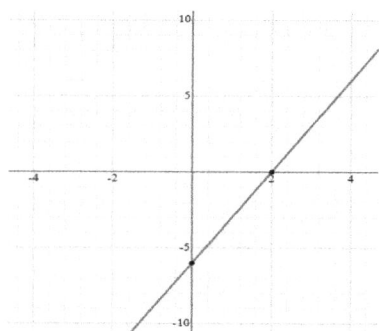

GED Subject Test – Mathematics

Writing Linear Equations

1) $y = 14x - 33$
2) $y = x + 9$
3) $y = -4x + 47$
4) $y = x - 4$
5) $y = 2x - 31$
6) $y = 9x - 43$
7) $y = -2x + 24$
8) $y = -3x + 1$
9) $y = -2x + 5$
10) $y = -2x + 6$
11) $y = 5x - 18$
12) $y = 14x - 122$
13) $y = 13x + 52$
14) $y = x + 10$
15) $y = -5x + 58$
16) $y = -5x + 5$
17) $y = 6x + 12$
18) $y = -11x - 4$
19) $y = -3x + 17$
20) $y = -5x - 11$
21) $y = -10x - 30$
22) $y = 8x + 7$

Graphing Linear Inequalities

1) $y > 4x - 5$

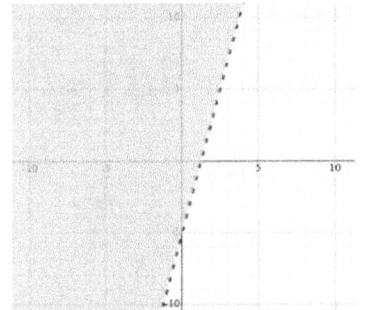

2) $y < 2x + 4$

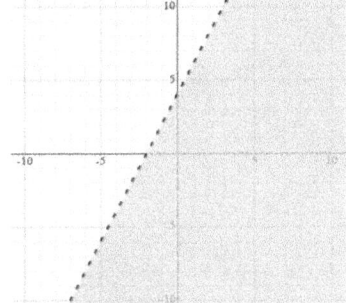

3) $y \leq -5x - 2$

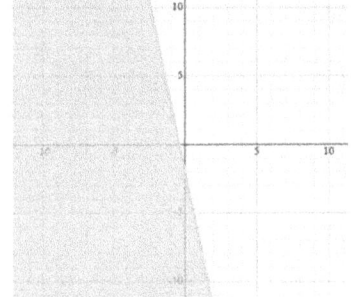

4) $4y \geq 12 + 4x$

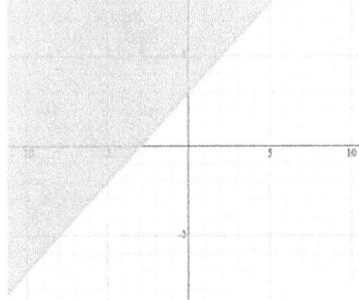

5) $-12y < 3x - 24$

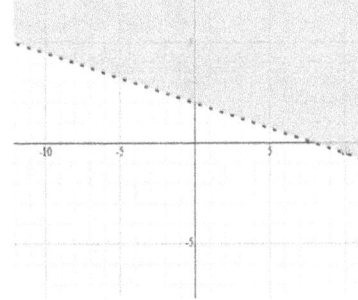

6) $5y \geq -15x + 10$

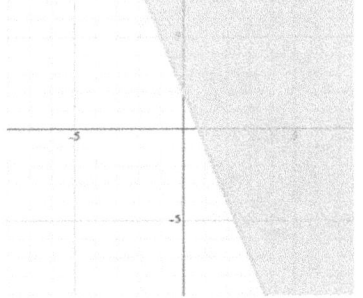

Finding Midpoint

1) $(-1, 0)$
2) $(4, 4)$
3) $(6, 4)$
4) $(-5, 1)$
5) $(8, -10)$
6) $(-3, -8)$
7) $(0, 10)$
8) $(-6, 4)$
9) $(6, -2)$
10) $(8, 6)$
11) $(-1, 6)$
12) $(4, 1)$
13) $(-4, 7.5)$
14) $(10, 2.5)$
15) $(2, 8)$
16) $(-4, 4)$
17) $(10, -10)$
18) $(4, -6)$

GED Subject Test – Mathematics

19) $(-1.5, -1)$ 22) $(-9, 9)$ 25) $(-9, 11)$
20) $(15, 5)$ 23) $(5, -4)$ 26) $(5, 3)$
21) $(4, -3)$ 24) $(16, 6)$ 27) $(14, 34)$

Finding Distance of Two Points

1) 15   9) 11   17) 15
2) 17   10) 10   18) 10
3) 2   11) 10   19) 18
4) 26   12) 25   20) 12
5) 9   13) $4\sqrt{10}$ 21) 6 *square units*
6) 29   14) 20   22) 32 *units*
7) 25   15) 6   23) 20 *units*
8) 13   16) 25

GED Subject Test – Mathematics

Chapter 8 : Polynomials

Topics that you'll practice in this chapter:

- ✓ Writing Polynomials in Standard Form
- ✓ Simplifying Polynomials
- ✓ Adding and Subtracting Polynomials
- ✓ Multiplying Monomials
- ✓ Multiplying and Dividing Monomials
- ✓ Multiplying a Polynomial and a Monomial
- ✓ Multiplying Binomials
- ✓ Factoring Trinomials
- ✓ Operations with Polynomials

Mathematics is the supreme judge; from its decisions there is no appeal. – Tobias Dantzig

GED Subject Test – Mathematics

Writing Polynomials in Standard Form

✏️ **Write each polynomial in standard form.**

1) $11x - 7x =$

2) $-5 + 19x - 19x =$

3) $6x^5 - 12x^3 =$

4) $12 + 17x^4 - 12 =$

5) $5x^2 + 4x - 9x^3 =$

6) $-3x^2 + 12x^5 =$

7) $5x + 8x^3 - 2x^8 =$

8) $-7x^3 + 4x - 9x^6 =$

9) $3x^2 + 22 - 6x =$

10) $3 - 4x + 9x^4 =$

11) $13x^2 + 28x - 8x^3 =$

12) $16 + 4x^2 - 2x^3 =$

13) $19x^2 - 9x + 9x^4 =$

14) $3x^4 - 7x^2 - 2x^3 =$

15) $-51 + 3x^2 - 8x^4 =$

16) $7x^2 - 8x^6 + 4x^4 - 15 =$

17) $6x^4 - 4x^5 + 16 - 3x^3 =$

18) $-2x^6 + 4x - 7x^2 - 5x =$

19) $11x^7 + 8x^5 - 5x^7 - 3x^2 =$

20) $2x^2 - 12x^5 + 8x^2 + 3x^6 =$

21) $4x^5 - 11x^7 - 6x^3 + 16x^5 =$

22) $6x^3 + 3x^5 + 34x^4 - 8x^5 =$

23) $3x(4x + 5 - 2x^2) =$

24) $12x(x^6 + 4x^3) =$

25) $5x(3x^2 + 6x + 4) =$

26) $7x(4 - 2x + 6x^5) =$

27) $3x(4x^4 - 4x^3 + 2) =$

28) $4x(2x^5 + 6x^2 - 3) =$

29) $5x(3x^4 + 4x^3 + 2x) =$

30) $2x(3x - 2x^3 + 4x^6) =$

WWW.MathNotion.Com

GED Subject Test – Mathematics

Simplifying Polynomials

✎ **Simplify each expression.**

1) $3(4x - 20) =$

2) $5x(3x - 4) =$

3) $6x(5x - 7) =$

4) $3x(7x + 5) =$

5) $5x(4x - 3) =$

6) $6x(8x + 2) =$

7) $(3x - 2)(x - 4) =$

8) $(x - 5)(2x + 6) =$

9) $(x - 3)(x - 7) =$

10) $(3x + 4)(3x - 4) =$

11) $(5x - 4)(5x - 2) =$

12) $6x^2 + 6x^2 - 8x^4 =$

13) $3x - 2x^2 + 5x^3 + 7 =$

14) $7x + 4x^2 - 10x^3 =$

15) $12x^2 + 5x^5 - 6x^3 =$

16) $-5x^2 + 4x^6 + 6x^8 =$

17) $-12x^3 + 10x^5 - 4x^6 + 4x =$

18) $11 - 7x^2 + 4x^2 - 16x^3 + 11 =$

19) $2x^2 - 9x + 4x^3 + 15x - 10x =$

20) $13 - 7x^5 + 6x^5 - 4x^2 + 5 =$

21) $-5x^8 + x^6 - 14x^3 + 5x^8 =$

22) $(7x^4 - 4) + (7x^4 - 2x^4) =$

23) $3(3x^4 - 4x^3 - 6x^4) =$

24) $-5(x^9 + 8) - 5(10 - x^9) =$

25) $8x^3 - 9x^4 - 2x + 19 - 8x^3 =$

26) $11 - 8x^3 + 6x^3 - 7x^5 + 6 =$

27) $(5x^3 - 4x) - (6x - 2 - 6x^3) =$

28) $4x^2 - 5x^4 - x(3x^3 + 2x) =$

29) $6x + 6x^5 - 10 - 4(x^5 - 3) =$

30) $4 - 3x^4 + (6x^5 - 2x^4 + 5x^5) =$

31) $-(x^5 + 4) - 8(3 + x^5) =$

32) $(4x^3 - 3x) - (3x - 5x^3) =$

WWW.MathNotion.Com

GED Subject Test – Mathematics

Adding and Subtracting Polynomials

✎ **Add or subtract expressions.**

1) $(-2x^2 - 3) + (3x^2 + 4) =$

2) $(4x^3 + 6) - (7 - 2x^3) =$

3) $(4x^5 + 5x^2) - (2x^5 + 15) =$

4) $(6x^3 - 2x^2) + (5x^2 - 4x) =$

5) $(10x^4 + 28x) - (34x^4 + 6) =$

6) $(7x^2 - 3) + (7x^2 + 3) =$

7) $(9x^2 + 4) - (10 - 5x^2) =$

8) $(6x^2 + x^5) - (x^5 + 4) =$

9) $(4x^3 - x) + (3x - 7x^3) =$

10) $(11x + 10) - (8x + 10) =$

11) $(15x^3 - 3x) - (3x - 4x^3) =$

12) $(4x - x^5) - (6x^5 + 8x) =$

13) $(2x^2 - 7x^7) - (4x^7 - 6x) =$

14) $(3x^2 - 5) + (8x^2 + 4x^5) =$

15) $(9x^4 + 5x^5) - (x^5 - 9x^4) =$

16) $(-4x^3 - 2x) + (9x - 5x^3) =$

17) $(4x - 3x^2) - (148x^2 + x) =$

18) $(5x - 8x^4) - (3x^4 - 4x^2) =$

19) $(8x^4 - 4) + (2x^4 - 3x^2) =$

20) $(5x^6 + 7x^3) - (x^3 - 5x^6) =$

21) $(-2x^2 + 20x^5 + 5x^4) + (12x^4 + 8x^5 + 24x^2) =$

22) $(7x^4 - 9x^7 - 6x) - (-3x^4 - 9x^7 + 6x) =$

23) $(14x + 12x^4 - 18x^6) + (20x^4 + 18x^6 - 10x) =$

24) $(5x^8 - 6x^6 - 4x) - (5x^3 + 9x^6 - 7x) =$

25) $(11x^2 - 6x^4 - 3x) - (-4x^2 - 12x^4 + 9x) =$

26) $(-5x^9 + 14x^3 + 3x^7) + (10x^7 + 26x^3 + 3x^9) =$

GED Subject Test – Mathematics

Multiplying Monomials

✎ **Simplify each expression.**

1) $6u^8 \times (-u^2) =$

2) $(-5p^8) \times (-2p^3) =$

3) $4xy^3z^5 \times 3z^4 =$

4) $3u^5t \times 8ut^4 =$

5) $(-5a^2) \times (-7a^3b^6) =$

6) $-3a^4b^3 \times 6a^2b =$

7) $13xy^5 \times x^4y^4 =$

8) $6p^4q^3 \times (-8pq^6) =$

9) $8s^4t^3 \times 4st^3 =$

10) $(-6x^4y^3) \times 6x^2y =$

11) $3xy^7z \times 12z^3 =$

12) $24xy \times x^2y =$

13) $13pq^4 \times (-3p^2q) =$

14) $13s^3t^4 \times st^4 =$

15) $11p^5 \times (-6p^3) =$

16) $(-8p^3q^5r) \times 3pq^4r^6 =$

17) $(-4a^4) \times (-7a^3b) =$

18) $6u^6v^2 \times (-5u^3v^4) =$

19) $9u^5 \times (-3u) =$

20) $-6xy^5 \times 4x^2y =$

21) $13y^5z^3 \times (-y^3z) =$

22) $8a^4bc^3 \times 2abc^3 =$

23) $(-7p^5q^6) \times (-5p^4q^2) =$

24) $4u^5v^3 \times (-4u^7v^3) =$

25) $17y^4z^5 \times (-y^6z) =$

26) $(-5pq^3r^2) \times 8p^2q^4r =$

27) $3ab^5c^6 \times 5a^4bc^2 =$

28) $6x^3yz^2 \times 3x^2y^7z^3 =$

WWW.MathNotion.Com

GED Subject Test – Mathematics

Multiplying and Dividing Monomials

✎ **Simplify each expression.**

1) $(5x^5)(2x^2) =$

2) $(4x^4)(6x^2) =$

3) $(3x^4)(7x^4) =$

4) $(5x^6)(4x^2) =$

5) $(12x^4)(3x^6) =$

6) $(4yx^8)(8y^4x^3) =$

7) $(14x^4y)(x^3y^5) =$

8) $(-5x^3y^4)(2x^3y^5) =$

9) $(-6x^4y^2)(-3x^3y^5) =$

10) $(5x^3y)(-5x^2y^3) =$

11) $(6x^4y^3)(4x^3y^4) =$

12) $(4x^3y^2)(5x^2y^4) =$

13) $(12x^3y^6)(4x^4y^{10}) =$

14) $(15x^3y^5)(3x^4y^6) =$

15) $(7x^2y^7)(8x^6y^7) =$

16) $(-3x^3y^8)(7x^9y^4) =$

17) $\dfrac{5x^6y^6}{xy^4} =$

18) $\dfrac{19x^7y^5}{19x^6y} =$

19) $\dfrac{56x^4y^4}{8xy} =$

20) $\dfrac{81x^5y^6}{9x^4y^5} =$

21) $\dfrac{36x^7y^6}{9x^2y^3} =$

22) $\dfrac{48x^9y^7}{4x^4y^6} =$

23) $\dfrac{88x^{18}y^{12}}{11x^8y^9} =$

24) $\dfrac{30x^7y^6}{6x^8y^3} =$

25) $\dfrac{150x^7y^6}{30x^4y^6} =$

26) $\dfrac{-42x^{18}y^{14}}{6x^4y^9} =$

27) $\dfrac{-36x^7y^8}{9x^5y^8} =$

WWW.MathNotion.Com

GED Subject Test – Mathematics

Multiplying a Polynomial and a Monomial

✏ **Find each product.**

1) $x(2x + 4) =$

2) $6(4 - 2x) =$

3) $5x(4x + 2) =$

4) $x(-4x + 5) =$

5) $8x(2x - 2) =$

6) $6(2x - 4y) =$

7) $7x(5x - 5) =$

8) $3x(12x + 2y) =$

9) $4x(x + 6y) =$

10) $11x(3x + 4y) =$

11) $7x(3x + 2) =$

12) $10x(4x - 10y) =$

13) $9x(3x - 2y) =$

14) $7x(x - 4y + 6) =$

15) $8x(2x^2 + 5y^2) =$

16) $12x(2x + 3y) =$

17) $4(2x^4 - 4y^4) =$

18) $4x(-3x^2y + 4y) =$

19) $-4(5x^3 - 2xy + 4) =$

20) $4(x^2 - 5xy - 6) =$

21) $8x(2x^3 - 5xy + 2x) =$

22) $-6x(-2x^3 - 6x + 2xy) =$

23) $3(2x^2 + xy - 9y^2) =$

24) $4x(5x^3 - 3x + 7) =$

25) $6(3x^{22} - 2x - 5) =$

26) $x^2(-2x^3 + 4x + 3) =$

27) $x^2(4x^3 + 10 - 2x) =$

28) $4x^4(3x^3 - 2x + 5) =$

29) $2x^2(4x^4 - 5xy + 7y^3) =$

30) $5x^2(5x^4 - 3x + 9) =$

31) $7x^2(6x^2 + 3x - 6) =$

32) $4x(x^3 - 4xy + 2y^2) =$

WWW.MathNotion.Com

GED Subject Test – Mathematics

Multiplying Binomials

✎ **Find each product.**

1) $(x + 3)(x + 6) =$

2) $(x - 4)(x + 3) =$

3) $(x - 3)(x - 8) =$

4) $(x + 8)(x + 9) =$

5) $(x - 2)(x - 12) =$

6) $(x + 5)(x + 5) =$

7) $(x - 6)(x + 7) =$

8) $(x - 8)(x - 3) =$

9) $(x + 7)(x + 12) =$

10) $(x - 4)(x + 8) =$

11) $(x + 8)(x + 8) =$

12) $(x + 2)(x + 7) =$

13) $(x - 6)(x + 6) =$

14) $(x - 5)(x + 5) =$

15) $(x + 11)(x + 11) =$

16) $(x + 6)(x + 9) =$

17) $(x - 2)(x + 2) =$

18) $(x - 4)(x + 7) =$

19) $(3x + 5)(x + 6) =$

20) $(5x - 6)(4x + 8) =$

21) $(x - 7)(3x + 7) =$

22) $(x - 9)(x - 4) =$

23) $(x - 12)(x + 2) =$

24) $(2x - 4)(5x + 4) =$

25) $(3x - 8)(x + 8) =$

26) $(7x - 2)(6x + 3) =$

27) $(4x + 5)(3x + 5) =$

28) $(7x - 4)(9x + 4) =$

29) $(x + 2)(2x - 8) =$

30) $(5x - 4)(5x + 4) =$

31) $(3x + 2)(3x - 7) =$

32) $(x^2 + 8)(x^2 - 8) =$

WWW.MathNotion.Com

GED Subject Test – Mathematics

Factoring Trinomials

✎ **Factor each trinomial.**

1) $x^2 + 8x + 12 =$

2) $x^2 - 6x + 5 =$

3) $x^2 + 15x + 36 =$

4) $x^2 - 12x + 35 =$

5) $x^2 - 11x + 18 =$

6) $x^2 - 9x + 18 =$

7) $x^2 + 18x + 72 =$

8) $x^2 - x - 72 =$

9) $x^2 + 4x - 21 =$

10) $x^2 - 13x + 22 =$

11) $x^2 + 2x - 24 =$

12) $x^2 - 3x - 40 =$

13) $x^2 - 3x - 70 =$

14) $x^2 + 26x + 169 =$

15) $4x^2 - 7x - 15 =$

16) $x^2 - 14x + 33 =$

17) $10x^2 + 5x - 15 =$

18) $6x^2 - 4x - 42 =$

19) $x^2 + 12x + 36 =$

20) $5x^2 + 17x - 12 =$

✎ **Calculate each problem.**

21) The area of a rectangle is $x^2 - x - 56$. If the width of rectangle is $x + 7$, what is its length? _____

22) The area of a parallelogram is $4x^2 + 17x - 15$ and its height is $x + 5$. What is the base of the parallelogram? _____

23) The area of a rectangle is $6x^2 - 22x + 12$. If the width of the rectangle is $3x - 2$, what is its length? _____

GED Subject Test – Mathematics

Operations with Polynomials

✎ **Find each product.**

1) $4(5x + 3) =$ _____

2) $8(2x + 6) =$ _____

3) $2(5x - 2) =$ _____

4) $-4(7x - 3) =$ _____

5) $3x^2(9x + 1) =$ _____

6) $4x^6(7x - 9) =$ _____

7) $3x^4(-7x + 3) =$ _____

8) $-8x^4(5x - 8) =$ _____

9) $7(x^2 + 5x - 3) =$ _____

10) $9(5x^2 - 7x + 5) =$ _____

11) $3(3x^2 + 3x + 2) =$ _____

12) $5x(3x^2 + 5x + 8) =$ _____

13) $(5x + 7)(3x - 3) =$ _____

14) $(9x + 3)(3x - 5) =$ _____

15) $(6x + 3)(4x - 2) =$ _____

16) $(7x - 2)(3x + 5) =$ _____

✎ **Calculate each problem.**

17) The measures of two sides of a triangle are $(2x + 5y)$ and $(6x - 3y)$. If the perimeter of the triangle is $(13x + 4y)$, what is the measure of the third side? _____

18) The height of a triangle is $(8x + 5)$ and its base is $(4x - 3)$. What is the area of the triangle? _____

19) One side of a square is $(6x + 2)$. What is the area of the square? _____

20) The length of a rectangle is $(5x - 8y)$ and its width is $(15x + 8y)$. What is the perimeter of the rectangle? _____

21) The side of a cube measures $(x + 2)$. What is the volume of the cube? _____

22) If the perimeter of a rectangle is $(28x + 6y)$ and its width is $(5x + 2y)$, what is the length of the rectangle? _____

GED Subject Test – Mathematics

Answers of Worksheets

Writing Polynomials in Standard Form

1) $4x$
2) -5
3) $6x^5 - 12x^3$
4) $14x^4$
5) $-9x^3 + 5x^2 + 4x$
6) $12x^5 - 3x^2$
7) $-2x^8 + 8x^3 + 5x$
8) $-9x^6 - 7x^3 + 4x$
9) $3x^2 - 6x + 22$
10) $9x^4 - 4x + 3$
11) $-8x^3 + 13x^2 + 28x$
12) $-2x^3 + 4x^2 + 16$
13) $9x^4 + 19x^2 - 9x$
14) $3x^4 - 2x^3 - 7x^2$
15) $-8x^4 + 3x^2 - 51$
16) $-8x^6 + 4x^4 + 7x^2 - 15$
17) $-4x^5 + 6x^4 - 3x^3 + 16$
18) $-2x^6 - 7x^2 - x$
19) $6x^7 + 8x^5 - 3x^2$
20) $3x^6 - 12x^5 + 10x^2$
21) $-11x^7 + 20x^5 - 6x^3$
22) $-5x^5 + 34x^4 + 6x^3$
23) $-6x^3 + 12x^2 + 15x$
24) $12x^7 + 48x^4$
25) $15x^3 + 30x^2 + 20x$
26) $42x^6 - 14x^2 + 28x$
27) $12x^5 - 12x^4 + 6x$
28) $8x^6 + 24x^3 - 12x$
29) $15x^5 + 20x^4 + 10x^2$
30) $8x^7 - 4x^4 + 6x^2$

Simplifying Polynomials

1) $12x - 60$
2) $15x^2 - 20x$
3) $30x^2 - 42x$
4) $21x^2 + 15x$
5) $20x^2 - 15x$
6) $48x^2 + 12x$
7) $3x^2 - 14x + 8$
8) $2x^2 - 4x - 30$
9) $x^2 - 10x + 21$
10) $9x^2 - 16$
11) $25x^2 - 30x + 8$
12) $-8x^4 + 12x^2$
13) $5x^3 - 2x^2 + 3x + 7$
14) $-10x^3 + 4x^2 + 7x$
15) $5x^5 - 6x^3 + 12x^2$
16) $6x^8 + 4x^6 - 5x^2$
17) $-4x^6 + 10x^5 - 12x^3 + 4x$
18) $-16x^3 - 3x^2 + 22$
19) $4x^3 + 2x^2 - 4x$
20) $-x^5 - 4x^2 + 18$
21) $x^6 - 14x^3$
22) $12x^4 - 4$
23) $-9x^4 - 12x^3$
24) -90

WWW.MathNotion.Com

GED Subject Test – Mathematics

25) $-9x^4 - 2x + 19$
26) $-7x^5 - 2x^3 + 17$
27) $11x^3 - 10x + 2$
28) $-8x^4 + 2x^2$

29) $2x^5 + 6x + 2$
30) $11x^5 - 5x^4 + 4$
31) $-9x^5 - 28$
32) $9x^3 - 6x$

Adding and Subtracting Polynomials

1) $x^2 + 1$
2) $6x^3 - 1$
3) $2x^5 + 5x^2 - 15$
4) $6x^3 + 3x^2 - 4x$
5) $-24x^4 + 28x - 6$
6) $14x^2$
7) $14x^2 - 6$
8) $6x^2 - 4$
9) $-3x^3 + 2x$

10) $3x$
11) $19x^3 - 6x$
12) $-7x^5 - 4x$
13) $-11x^7 + 2x^2 + 6x$
14) $4x^5 + 11x^2 - 5$
15) $4x^5 + 18x^4$
16) $-9x^3 + 7x$
17) $-151x^2 + 3x$
18) $-11x^4 + 4x^2 + 5x$

19) $10x^4 - 3x^2 - 4$
20) $10x^6 + 6x^3$
21) $28x^5 + 17x^4 + 22x^2$
22) $10x^4 - 12x$
23) $32x^4 + 4x$
24) $5x^8 - 15x^6 - 5x^3 + 3x$
25) $6x^4 + 15x^2 - 12x$
26) $-2x^9 + 13x^7 + 40x^3$

Multiplying Monomials

1) $-6u^{10}$
2) $10p^{11}$
3) $12xy^3z^9$
4) $24u^6t^5$
5) $35a^5b^6$
6) $-18a^6b^4$
7) $13x^5y^9$
8) $-48p^5q^9$
9) $32s^5t^6$
10) $-36x^6y^4$

11) $36xy^7z^4$
12) $24px^3y^2$
13) $-39p^3q^5$
14) $13s^4t^8$
15) $-66p^8$
16) $-24p^4q^9r^7$
17) $28a^7b$
18) $-30u^9v^6$
19) $-27u^6$
20) $-24x^3y^6$

21) $-13y^8z^4$
22) $16a^5b^2c^6$
23) $35p^9q^8$
24) $-16u^{12}v^6$
25) $-17y^{10}z^6$
26) $-40p^3q^7r^3$
27) $15a^5b^6c^8$
28) $18x^5y^8z^5$

Multiplying and Dividing Monomials

1) $10x^7$
2) $24x^6$
3) $21x^8$
4) $20x^8$

5) $36x^{10}$
6) $32x^{11}y^5$
7) $14x^7y^6$
8) $-10x^6y^9$

9) $18x^7y^7$
10) $-25x^5y^4$
11) $24x^7y^7$
12) $20x^5y^6$

GED Subject Test – Mathematics

13) $48x^7y^{16}$
14) $45x^7y^{11}$
15) $56x^8y^{14}$
16) $-21x^{12}y^{12}$
17) $5x^5y^2$

18) xy^4
19) $7x^3y^3$
20) $9xy$
21) $4x^5y^3$
22) $12x^5y$

23) $8x^{10}y^3$
24) $5x^{-1}y^3$
25) $5x^3$
26) $-7x^{14}y^5$
27) $-4x^2$

Multiplying a Polynomial and a Monomial

1) $2x^2 + 4x$
2) $-12x + 24$
3) $20x^2 + 10x$
4) $-4x^2 + 5x$
5) $16x^2 - 16x$
6) $12x - 24y$
7) $35x^2 - 35x$
8) $36x^2 + 6xy$
9) $4x^2 + 24xy$
10) $33x^2 + 44xy$
11) $21x^2 + 14x$
12) $40x^2 - 100xy$
13) $27x^2 - 18xy$
14) $7x^2 - 28xy + 42x$
15) $16x^3 + 40xy^2$
16) $24x^2 + 36xy$

17) $8x^4 - 16y^4$
18) $-12x^3y + 16xy$
19) $-20x^3 + 8xy - 16$
20) $4x^2 - 20xy - 24$
21) $16x^4 - 40x^2y + 16x^2$
22) $12x^4 + 36x^2 - 12x^2y$
23) $6x^2 + 3xy - 27y^2$
24) $20x^4 - 12x^2 + 28x$
25) $18x^{22} - 12x - 30$
26) $-2x^5 + 4x^3 + 3x^2$
27) $4x^5 - 2x^3 + 10x^2$
28) $12x^7 - 8x^5 + 20x^4$
29) $8x^6 - 10x^3y + 14x^2y^3$
30) $25x^6 - 15x^3 + 45x^2$
31) $42x^4 + 21x^3 - 42x^2$
32) $4x^4 - 16x^2y + 8xy^2$

Multiplying Binomials

1) $x^2 + 9x + 18$
2) $x^2 - x - 12$
3) $x^2 - 11x + 24$
4) $x^2 + 17x + 72$
5) $x^2 - 14x + 24$
6) $x^2 + 10x + 25$
7) $x^2 + x - 42$

8) $x^2 - 11x + 24$
9) $x^2 + 19x + 84$
10) $x^2 + 4x - 32$
11) $x^2 + 16x + 64$
12) $x^2 + 9x + 14$
13) $x^2 - 36$
14) $x^2 - 25$

GED Subject Test – Mathematics

15) $x^2 + 22x + 121$
16) $x^2 + 15x + 54$
17) $x^2 - 4$
18) $x^2 + 3x - 28$
19) $3x^2 + 23x + 30$
20) $20x^2 + 16x - 48$
21) $3x^2 - 14x - 49$
22) $x^2 - 13x + 36$
23) $x^2 - 10x - 24$

24) $10x^2 - 12x - 16$
25) $3x^2 + 16x - 64$
26) $42x^2 + 9x - 6$
27) $12x^2 + 35x + 25$
28) $63x^2 - 8x - 16$
29) $2x^2 - 4x - 16$
30) $25x^2 - 16$
31) $9x^2 - 15x - 14$
32) $x^4 - 64$

Factoring Trinomials

1) $(x + 6)(x + 2)$
2) $(x - 5)(x - 1)$
3) $(x + 12)(x + 3)$
4) $(x - 5)(x - 7)$
5) $(x - 2)(x - 9)$
6) $(x - 6)(x - 3)$
7) $(x + 6)(x + 12)$
8) $(x + 8)(x - 9)$

9) $(x - 3)(x + 7)$
10) $(x - 11)(x - 2)$
11) $(x - 4)(x + 6)$
12) $(x - 8)(x + 5)$
13) $(x + 7)(x - 10)$
14) $(x + 13)(x + 13)$
15) $(4x + 5)(x - 3)$
16) $(x - 11)(x - 3)$

17) $(5x - 5)(2x + 3)$
18) $(2x - 6)(3x + 7)$
19) $(x + 6)(x + 6)$
20) $(5x - 3)(x + 4)$
21) $(x - 8)$
22) $(4x - 3)$
23) $(2x - 6)$

Operations with Polynomials

1) $20x + 12$
2) $16x + 48$
3) $10x - 4$
4) $-28x + 12$
5) $27x^3 + 3x^2$
6) $28x^7 - 36x^6$
7) $-21x^5 + 9x^4$
8) $-40x^5 + 64x^4$

9) $7x^2 + 35x - 21$
10) $45x^2 - 63x + 45$
11) $9x^2 + 9x + 6$
12) $15x^3 + 25x^2 + 40x$
13) $15x^2 + 6x - 21$
14) $27x^2 - 36x - 15$
15) $24x^2 - 6$
16) $21x^2 + 29x - 10$

17) $(5x + 2y)$
18) $16x^2 - 2x - \frac{15}{2}$
19) $36x^2 + 24x + 4$
20) $40x$
21) $x^3 + 6x^2 + 12x + 8$
22) $(9x + y)$

WWW.MathNotion.Com

GED Subject Test – Mathematics

Chapter 9 : Functions Operations and Quadratic

Topics that you'll practice in this chapter:

- ✓ Evaluating Function
- ✓ Adding and Subtracting Functions
- ✓ Multiplying and Dividing Functions
- ✓ Composition of Functions
- ✓ Quadratic Equation
- ✓ Solving Quadratic Equations
- ✓ Quadratic Formula and the Discriminant
- ✓ Graphing Quadratic Functions

It's fine to work on any problem, so long as it generates interesting mathematics along the way – even if you don't solve it at the end of the day." – Andrew Wiles

GED Subject Test – Mathematics

Evaluating Function

✎ **Write each of following in function notation.**

1) $h = -8x + 3$

2) $k = 2a - 14$

3) $d = 11t$

4) $y = \frac{5}{12}x - \frac{7}{12}$

5) $m = 24n - 210$

6) $c = p^2 - 5p + 10$

✎ **Evaluate each function.**

7) $f(x) = 2x - 7$, find $f(-3)$

8) $g(x) = \frac{1}{9}x + 12$, find $f(18)$

9) $h(x) = -4x + 9$, find $f(3)$

10) $f(x) = -x + 19$, find $f(-3)$

11) $f(a) = 7a - 12$, find $f(3)$

12) $h(x) = 14 - 3x$, find $f(-4)$

13) $g(n) = 6n - 10$, find $f(2)$

14) $f(x) = -11x - 4$, find $f(-1)$

15) $k(n) = -20 - 3.5n$, find $f(2)$

16) $f(x) = -0.7x + 3.3$, find $f(-7)$

17) $g(n) = \frac{11n+8}{n}$, find $g(2)$

18) $g(n) = \sqrt{3n} + 12$, find $g(3)$

19) $h(x) = x^{-2} - 7$, find $h(\frac{1}{9})$

20) $h(n) = n^{-3} + 11$, find $h(\frac{1}{4})$

21) $h(n) = n^3 - 2$, find $h(\frac{1}{2})$

22) $h(n) = n^2 - 4$, find $h(-\frac{1}{3})$

23) $h(n) = 4n^2 - 13$, find $h(-5)$

24) $h(n) = -2n^3 - 6n$, find $h(2)$

25) $g(n) = \sqrt{16n^2} - \sqrt{n}$, find $g(4)$

26) $h(a) = \frac{-14a+9}{3a}$, find $h(-b)$

27) $k(a) = 12a - 14$, find $k(a - 3)$

28) $h(x) = \frac{1}{9}x + 18$, find $h(-18x)$

29) $h(x) = 8x^2 + 16$, find $h(\frac{x}{2})$

30) $h(x) = x^4 - 20$, find $h(-2x)$

WWW.MathNotion.Com

GED Subject Test – Mathematics

Adding and Subtracting Functions

✏️ **Perform the indicated operation.**

1) $f(x) = 2x + 3$
   $g(x) = x + 7$
   Find $(f - g)(2)$

2) $g(a) = -5a - 8$
   $f(a) = -3a - 5$
   Find $(g - f)(-2)$

3) $h(t) = 4t + 3$
   $g(t) = 4t + 7$
   Find $(h - g)(t)$

4) $g(a) = -6a - 10$
   $f(a) = 3a^2 + 9$
   Find $(g - f)(x)$

5) $g(x) = \frac{5}{6}x - 23$
   $h(x) = \frac{5}{12}x + 25$
   Find $g(12) - h(12)$

6) $h(x) = \sqrt{3x} - 2$
   $g(x) = \sqrt{3x} + 5$
   Find $(h + g)(12)$

7) $f(x) = x^{-1}$
   $g(x) = x^2 + \frac{5}{x}$
   Find $(f - g)(-3)$

8) $h(n) = n^2 + 2$
   $g(n) = -4n + 6$
   Find $(h - g)(2a)$

9) $g(x) = -2x^2 - 5 - 4x$
   $f(x) = 7 + 2x$
   Find $(g - f)(3x)$

10) $g(t) = 11t - 4$
    $f(t) = -2t^2 + 5$
    Find $(g + f)(-t)$

11) $f(x) = 8x + 9$
    $g(x) = -5x^2 + 3x$
    Find $(f - g)(-x^2)$

12) $f(x) = -3x^4 - 5x$
    $g(x) = 2x^4 + 5x + 22$
    Find $(f + g)(3x^2)$

WWW.MathNotion.Com

GED Subject Test – Mathematics

Multiplying and Dividing Functions

✎ **Perform the indicated operation.**

1) $g(x) = -2x - 1$
   $f(x) = 4x + 3$
   Find $(g.f)(2)$

2) $f(x) = 5x$
   $h(x) = -2x + 3$
   Find $(f.h)(-2)$

3) $g(a) = 5a - 2$
   $h(a) = 2a - 3$
   Find $(g.h)(-3)$

4) $f(x) = 2x - 7$
   $h(x) = x - 5$
   Find $\left(\frac{f}{h}\right)(4)$

5) $f(x) = 8a^2$
   $g(x) = 3 + 2a$
   Find $\left(\frac{f}{g}\right)(2)$

6) $g(a) = \sqrt{4a} + 2$
   $f(a) = (-a)^4 + 1$
   Find $\left(\frac{g}{f}\right)(1)$

7) $g(t) = t^3 + 1$
   $h(t) = 5t - 2$
   Find $(g.h)(-2)$

8) $g(n) = n^2 + 2n - 4$
   $h(n) = -5n + 3$
   Find $(g.h)(1)$

9) $g(a) = (a - 3)^2$
   $f(a) = a^2 + 4$
   Find $\left(\frac{g}{f}\right)(3)$

10) $g(x) = -3x^2 + \frac{4}{5}x + 9$
    $f(x) = x^2 - 24$
    Find $\left(\frac{g}{f}\right)(5)$

11) $f(x) = 2x^3 - 5x^2 + 1$
    $g(x) = 3x - 1$
    Find $(f.g)(x)$

12) $f(x) = 5x - 2$
    $g(x) = x^3 - 2x$
    Find $(f.g)(x^2)$

WWW.MathNotion.Com

Composition of Functions

Using $f(x) = 2x - 5$ and $g(x) = -2x$, find:

1) $f(g(2)) =$

2) $f(g(-1)) =$

3) $g(f(-4)) =$

4) $g(f(5)) =$

5) $f(g(3)) =$

6) $g(f(0)) =$

Using $f(x) = -\frac{1}{4}x + \frac{3}{4}$ and $g(x) = 2x^2$, find:

7) $g(f(-2)) =$

8) $g(f(4)) =$

9) $g(g(1)) =$

10) $f(f(1)) =$

11) $g(f(-4)) =$

12) $f(g(x)) =$

Using $f(x) = -2x + 2$ and $g(x) = x + 1$, find:

13) $g(f(1)) =$

14) $f(f(0)) =$

15) $f(g(-1)) =$

16) $f(g(-3)) =$

17) $g(f(2)) =$

18) $f(g(x)) =$

Using $f(x) = \sqrt{x + 9}$ and $g(x) = x - 9$, find:

19) $f(g(9)) =$

20) $g(f(-9)) =$

21) $f(g(4)) =$

22) $f(f(7)) =$

23) $g(f(-5)) =$

24) $g(g(0)) =$

GED Subject Test – Mathematics

Quadratic Equation

✎ Multiply.

1) $(x - 4)(x + 6) = $ _____

2) $(x + 5)(x + 7) = $ _____

3) $(x - 6)(x + 8) = $ _____

4) $(x + 2)(x - 9) = $ _____

5) $(x - 7)(x - 8) = $ _____

6) $(3x + 2)(x - 3) = $ _____

7) $(4x - 3)(x + 2) = $ _____

8) $(4x - 5)(x + 1) = $ _____

9) $(7x + 1)(x - 6) = $ _____

10) $(5x + 1)(3x - 3) = $ _____

✎ Factor each expression.

11) $x^2 - 2x - 8 = $ _____

12) $x^2 + 8x + 15 = $ _____

13) $x^2 - 2x - 24 = $ _____

14) $x^2 - 10x + 21 = $ _____

15) $x^2 + 10x + 21 = $ _____

16) $4x^2 + 9x + 5 = $ _____

17) $5x^2 + 13x - 6 = $ _____

18) $5x^2 + 17x - 12 = $ _____

19) $2x^2 + 7x + 5 = $ _____

20) $9x^2 - 21x + 6 = $ _____

✎ Calculate each equation.

21) $(x + 6)(x - 3) = 0$

22) $(x + 1)(x + 8) = 0$

23) $(3x + 6)(x + 5) = 0$

24) $(2x - 2)(4x + 8) = 0$

25) $x^2 + x + 10 = 22$

26) $x^2 + 11x + 36 = 12$

27) $2x^2 + 9x + 9 = 5$

28) $x^2 + 3x - 24 = 4$

29) $5x^2 + 5x - 40 = 20$

30) $8x^2 + 8x = 48$

WWW.MathNotion.Com

GED Subject Test – Mathematics

Solving Quadratic Equations

✎ Solve each equation by factoring or using the quadratic formula.

1) $(x+9)(x-1)=0$

2) $(x+7)(x+6)=0$

3) $(x-8)(x+3)=0$

4) $(x-6)(x-4)=0$

5) $(x+2)(x+12)=0$

6) $(5x+4)(x+7)=0$

7) $(6x+1)(4x+5)=0$

8) $(2x+7)(x+8)=0$

9) $(x+6)(3x+15)=0$

10) $(12x+2)(x+8)=0$

11) $x^2=8x$

12) $x^2-16=0$

13) $3x^2+6=9x$

14) $-2x^2-8=10x$

15) $5x^2+40x=45$

16) $x^2+10x=24$

17) $x^2+6x=16$

18) $x^2+9x=-18$

19) $x^2+13x=-36$

20) $x^2+3x-15=5x$

21) $x^2+8x+7=-8$

22) $3x^2-11x=-9+x$

23) $10x^2+3=27x-15$

24) $7x^2-6x+8=8$

25) $2x^2-12=-3x+2$

26) $10x^2-26x-3=-15$

27) $3x^2+21=-16x+5$

28) $x^2+15x-10=-66$

29) $3x^2-8x-8=4+x$

30) $2x^2+6x-24=12$

31) $3x^2-33x+54=-18$

32) $-10x^2-15x-9=-9-27x^2$

WWW.MathNotion.Com

GED Subject Test – Mathematics

Quadratic Formula and the Discriminant

✎ Find the value of the discriminant of each quadratic equation.

1) $3x(x - 8) = 0$

2) $2x^2 + 6x - 4 = 0$

3) $x^2 + 6x + 7 = 0$

4) $x^2 - x + 3 = 0$

5) $x^2 + 4x - 3 = 0$

6) $2x^2 + 6x - 10 = 0$

7) $3x^2 + 7x + 5 = 0$

8) $x^2 - 6x - 4 = 0$

9) $2x^2 + 8x + 3 = 0$

10) $x^2 + 7x - 5 = 0$

11) $5x^2 + 2x - 3 = 0$

12) $-3x^2 - 11x + 4 = 0$

13) $-6x^2 - 12x + 8 = 0$

14) $-x^2 - 9x - 12 = 0$

15) $7x^2 - 6x - 10 = 0$

16) $-4x^2 - 2x + 8 = 0$

17) $5x^2 + 8x - 2 = 0$

18) $6x^2 - 4x = 0$

19) $3x^2 - 5x + 2 = 0$

20) $4x^2 + 9x + 3 = 0$

✎ Find the discriminant of each quadratic equation then state the number of real and imaginary solutions.

21) $-4x^2 - 16 = 16x$

22) $20x^2 = 20x - 5$

23) $-11x^2 - 19x = 26$

24) $22x^2 - 4x + 1 = 18x^2$

25) $-11x^2 = -15x + 8$

26) $3x^2 + 6x + 9 = 6$

27) $13x^2 - 5x - 12 = -26$

28) $-8x^2 - 32x - 25 = 7$

WWW.MathNotion.Com

GED Subject Test – Mathematics

Graphing Quadratic Functions

✎ Sketch the graph of each function. Identify the vertex and axis of symmetry.

1) $y = (x + 3)^2 + 2$

2) $y = (x - 3)^2 - 2$

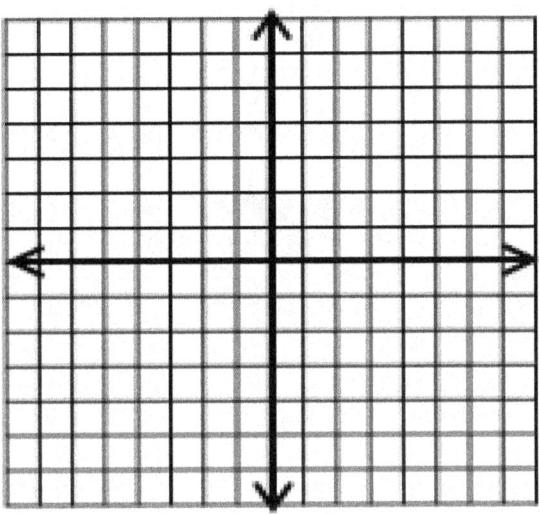
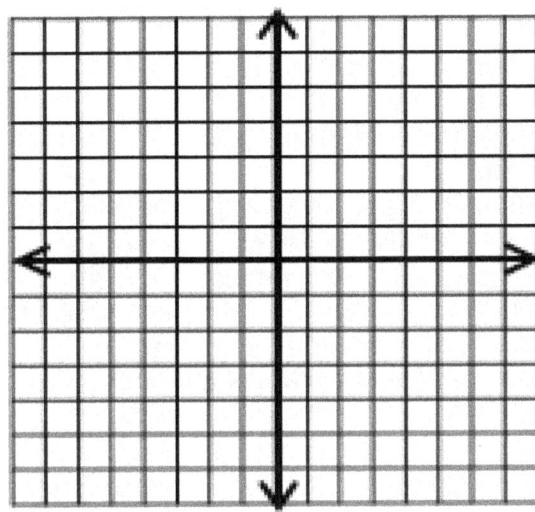

3) $y = 6 - (-x + 4)^2$

4) $y = -3x^2 - 6x + 9$

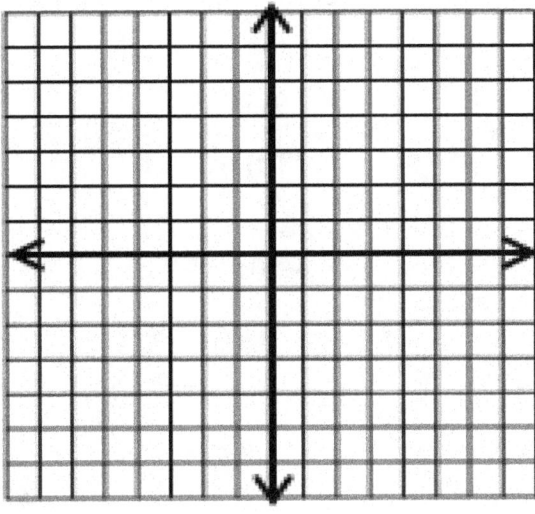
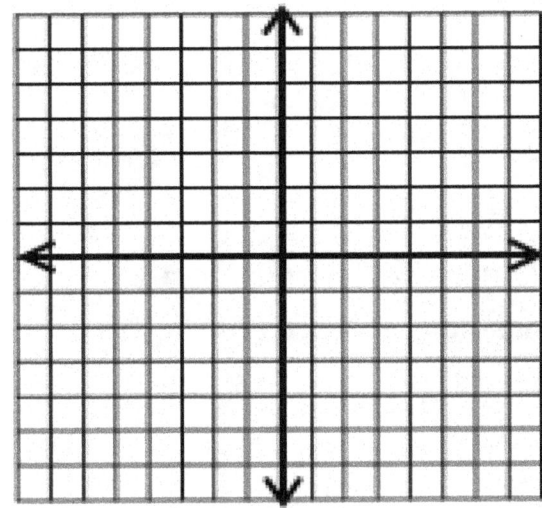

GED Subject Test – Mathematics

Answers of Worksheets

Evaluating Function

1) $h(x) = -8x + 3$
2) $k(a) = 2a - 14$
3) $d(t) = 11t$
4) $f(x) = \frac{5}{12}x - \frac{7}{12}$
5) $m(n) = 24n - 210$
6) $c(p) = p^2 - 5p + 10$
7) -13
8) 14
9) -3
10) 22
11) 9
12) 26
13) 2
14) 7
15) -27
16) 8.2
17) 15
18) 15
19) 74
20) 75
21) $-1\frac{7}{8}$
22) $-3\frac{8}{9}$
23) 87
24) -28
25) 14
26) $-\frac{14b+9}{3b}$
27) $12a - 50$
28) $-2x + 18$
29) $2x^2 + 16$
30) $16x^4 - 20$

Adding and Subtracting Functions

1) -2
2) 1
3) -4
4) $-3x^2 - 6x - 19$
5) -43
6) 15
7) $-7\frac{2}{3}$
8) $4a^2 + 8a - 4$
9) $-18x^2 - 18x - 12$
10) $-2t^2 - 11t + 1$
11) $5x^4 - 5x^2 + 9$
12) $-81x^8 + 22$

Multiplying and Dividing Functions

1) -55
2) -70
3) 153
4) -1
5) $4\frac{4}{7}$
6) 2
7) 84
8) 2
9) 0
10) -62
11) $6x^4 - 17x^3 + 5x^2 + 3x - 1$
12) $5x^8 - 2x^6 - 10x^4 + 4x^2$

Composition of Functions

1) -13
2) -1
3) 26
4) -10
5) -17
6) 10
7) $\frac{25}{8}$
8) $\frac{1}{8}$
9) 8
10) $\frac{5}{8}$
11) $\frac{49}{8}$
12) $-\frac{1}{2}(x^2 - \frac{3}{2})$
13) 1
14) -2
15) 2
16) 6
17) -1
18) $-2x$

WWW.MathNotion.Com

GED Subject Test – Mathematics

19) 3
20) −9
21) 2
22) $\sqrt{13}$
23) −7
24) −18

Quadratic Equations

1) $x^2 + 2x - 24$
2) $x^2 + 12x + 35$
3) $x^2 + 2x - 48$
4) $x^2 - 7x - 18$
5) $x^2 - 15x + 56$
6) $3x^2 - 7x - 6$
7) $4x^2 + 5x - 6$
8) $4x^2 - x - 5$
9) $7x^2 - 41x - 6$
10) $15x^2 - 12x - 3$
11) $(x - 4)(x + 2)$
12) $(x + 5)(x + 3)$
13) $(x - 6)(x + 4)$
14) $(x - 3)(x - 7)$
15) $(x + 3)(x + 7)$
16) $(4x + 5)(x + 1)$
17) $(5x - 2)(x + 3)$
18) $(5x - 3)(x + 4)$
19) $(2x + 5)(x + 1)$
20) $3(x - 2)(3x - 1)$
21) $x = -6, x = 3$
22) $x = -1, x = -8$
23) $x = -2, x = -5$
24) $x = 1, x = -2$
25) $x = 3, x = -4$
26) $x = -3, x = -8$
27) $x = -4, x = -\frac{1}{2}$
28) $x = 4, x = -7$
29) $x = 3, x = -4$
30) $x = -3, x = 2$

Solving quadratic equations

1) $\{-9, 1\}$
2) $\{-6, -7\}$
3) $\{8, -3\}$
4) $\{6, 4\}$
5) $\{-2, -12\}$
6) $\{-\frac{4}{5}, -7\}$
7) $\{-\frac{5}{4}, -\frac{1}{6}\}$
8) $\{-\frac{7}{2}, -8\}$
9) $\{-6, -5\}$
10) $\{-\frac{1}{6}, -8\}$
11) $\{8, 0\}$
12) $\{4, -4\}$
13) $\{2, 1\}$
14) $\{-4, -1\}$
15) $\{1, -9\}$
16) $\{2, -12\}$
17) $\{2, -8\}$
18) $\{-3, -6\}$
19) $\{-4, -9\}$
20) $\{5, -3\}$
21) $\{-5, -3\}$
22) $\{1, 3\}$
23) $\{\frac{6}{5}, \frac{3}{2}\}$
24) $\{\frac{6}{7}, 0\}$
25) $\{-\frac{7}{2}, 2\}$
26) $\{\frac{3}{5}, 2\}$
27) $\{-\frac{4}{3}, -4\}$
28) $\{-8, -7\}$
29) $\{4, -1\}$
30) $\{3, -6\}$
31) $\{3, 8\}$
32) $\{\frac{15}{17}, 0\}$

Quadratic formula and the discriminant

1) 576
2) 68
3) 8
4) −11
5) 28
6) 116
7) −11
8) 52
9) 40
10) 69
11) 64
12) 169
13) 336
14) 33
15) 316
16) 132
17) 104
18) 16
19) 1
20) 33
21) 0, one real solution
22) 0, one real solution
23) −783, no solution

WWW.MathNotion.Com

GED Subject Test – Mathematics

24) 0, one real solution 26) 0, one real solution 28) 0, one real solution
25) −127, no solution 27) −703, no solution

Graphing quadratic functions

1) $(-3, 2), x = -3$

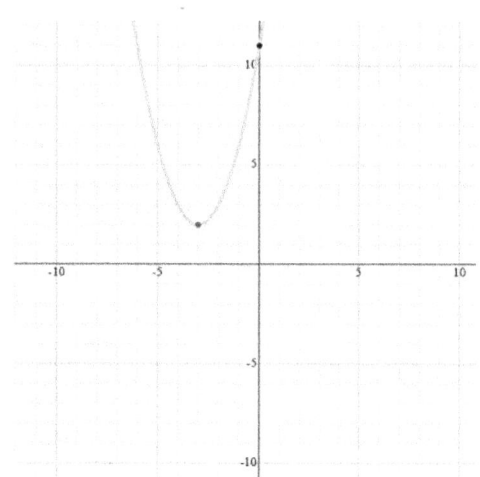

2) $(3, -2), x = 3$

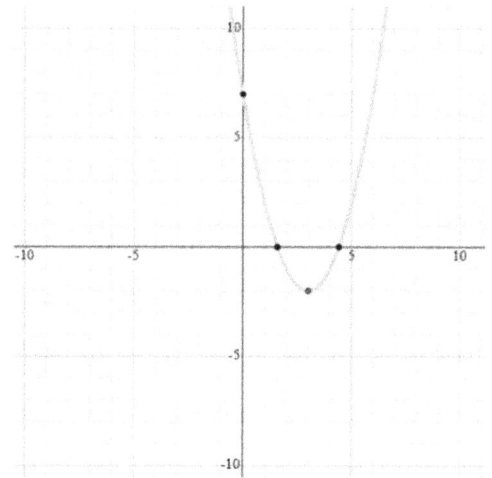

3) $(4, 6), x = 4$

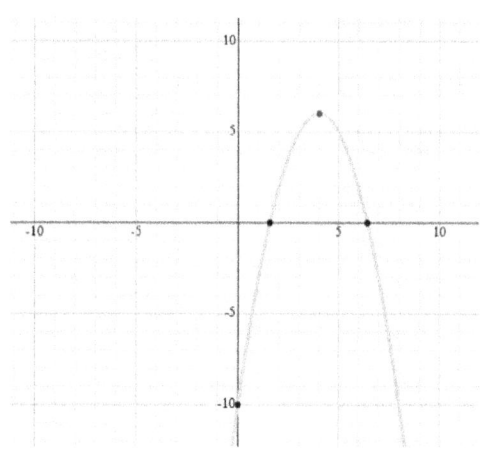

4) $(-1, 12), x = -1$

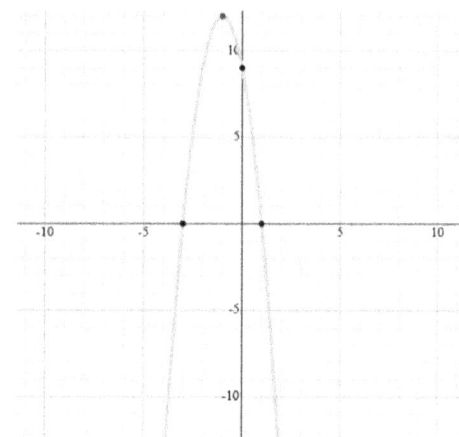

GED Subject Test – Mathematics

Chapter 10 :
Geometry and Solid Figures

Topics that you'll practice in this chapter:

- ✓ Angles
- ✓ Pythagorean Relationship
- ✓ Triangles
- ✓ Polygons
- ✓ Trapezoids
- ✓ Circles
- ✓ Cubes
- ✓ Rectangular Prism
- ✓ Cylinder
- ✓ Pyramids and Cone

Mathematics is, as it were, a sensuous logic, and relates to philosophy as do the arts, music, and plastic art to poetry. — K. Shegel

GED Subject Test – Mathematics

Angles

✎ **What is the value of x in the following figures?**

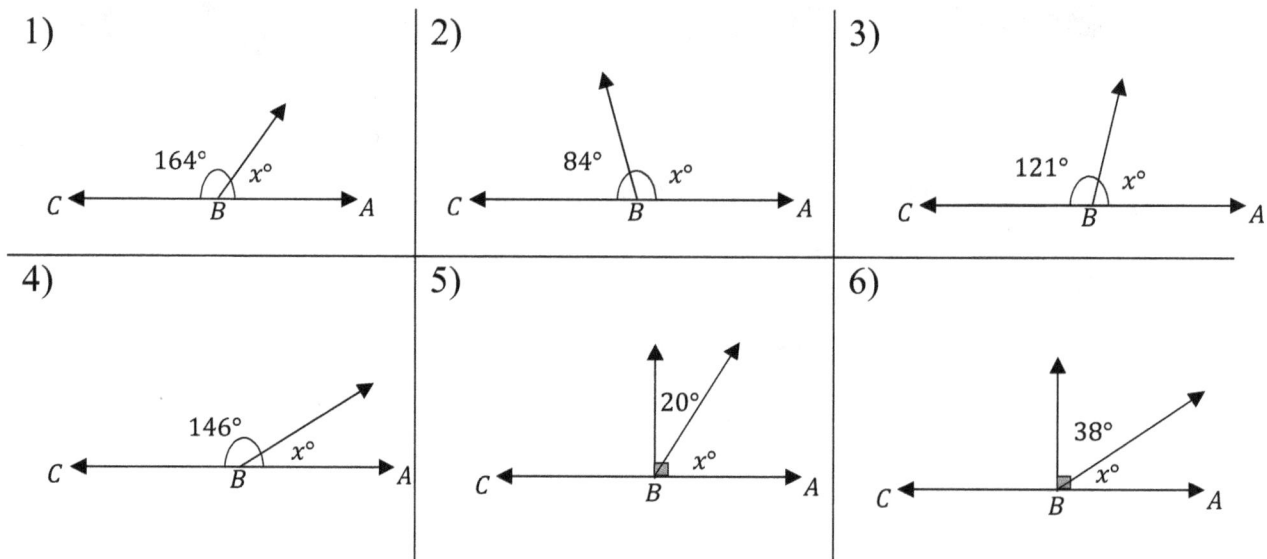

✎ **Calculate.**

7) Two supplement angles have equal measures. What is the measure of each angle? _____

8) The measure of an angle is seven fifth the measure of its supplement. What is the measure of the angle? _____

9) Two angles are complementary and the measure of one angle is 24 less than the other. What is the measure of the smaller angle? _____

10) Two angles are complementary. The measure of one angle is one fifth the measure of the other. What is the measure of the bigger angle? _____

11) Two supplementary angles are given. The measure of one angle is 40° less than the measure of the other. What does the smaller angle measure? _____

GED Subject Test – Mathematics

Pythagorean Relationship

 Do the following lengths form a right triangle?

1)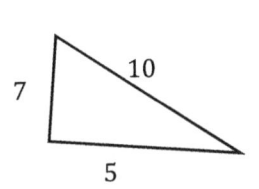

2)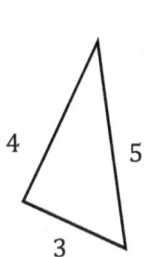

3)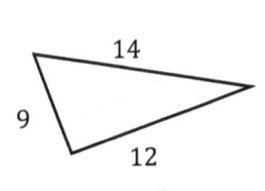

4)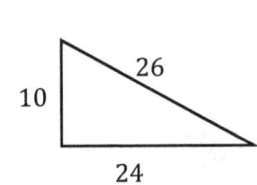

5)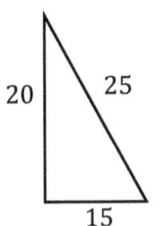

6)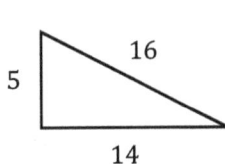

7)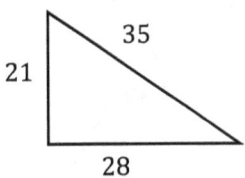

8)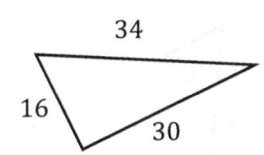

 Find the missing side?

9)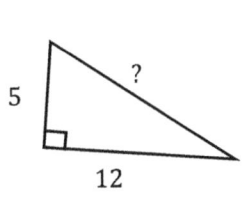

10)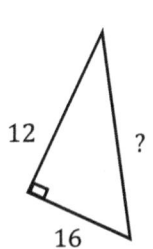

11)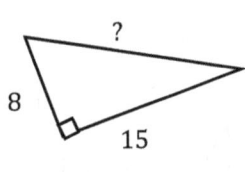

12)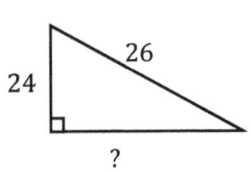

13)

14)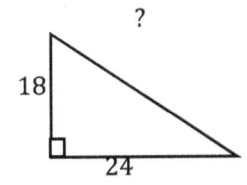

15)

16)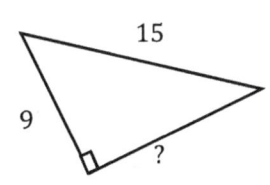

WWW.MathNotion.Com

GED Subject Test – Mathematics

Triangles

✏️ **Find the measure of the unknown angle in each triangle.**

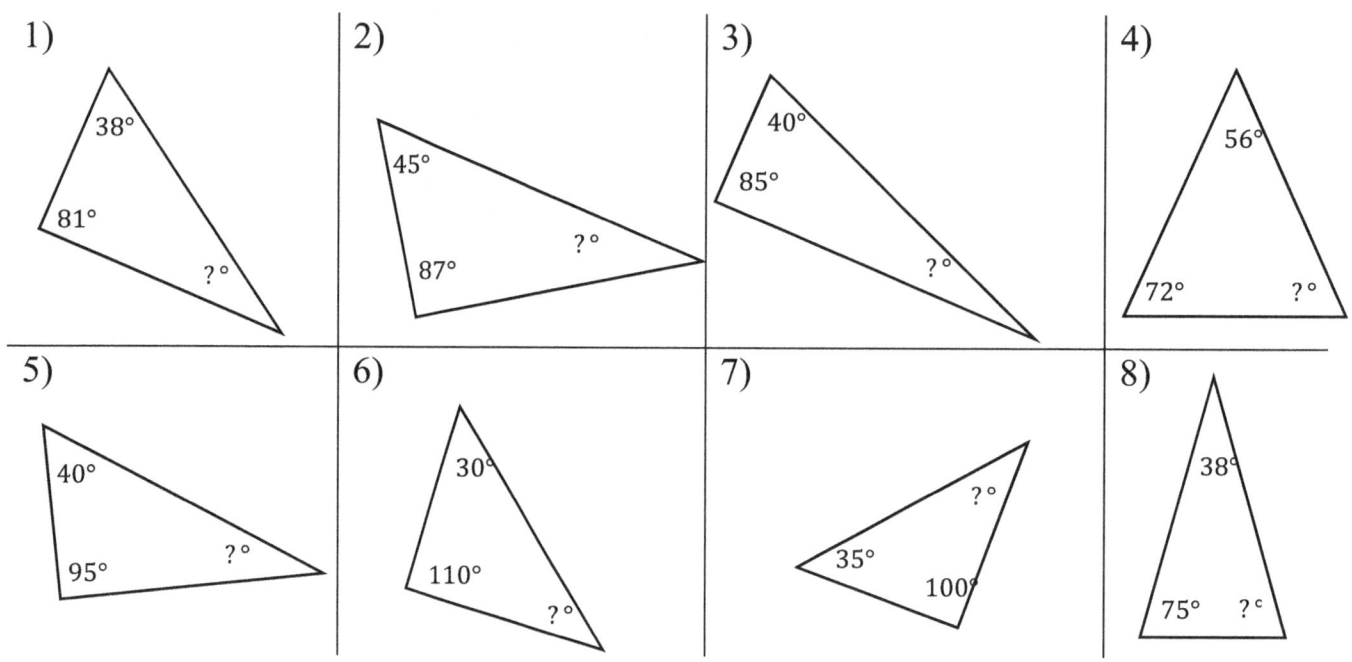

✏️ **Find area of each triangle.**

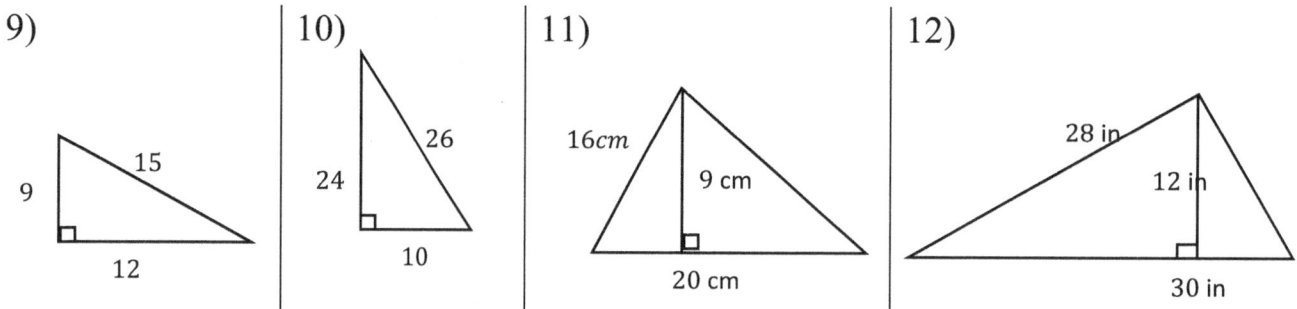

WWW.MathNotion.Com   126

GED Subject Test – Mathematics

Polygons

✍ **Find the perimeter of each shape.**

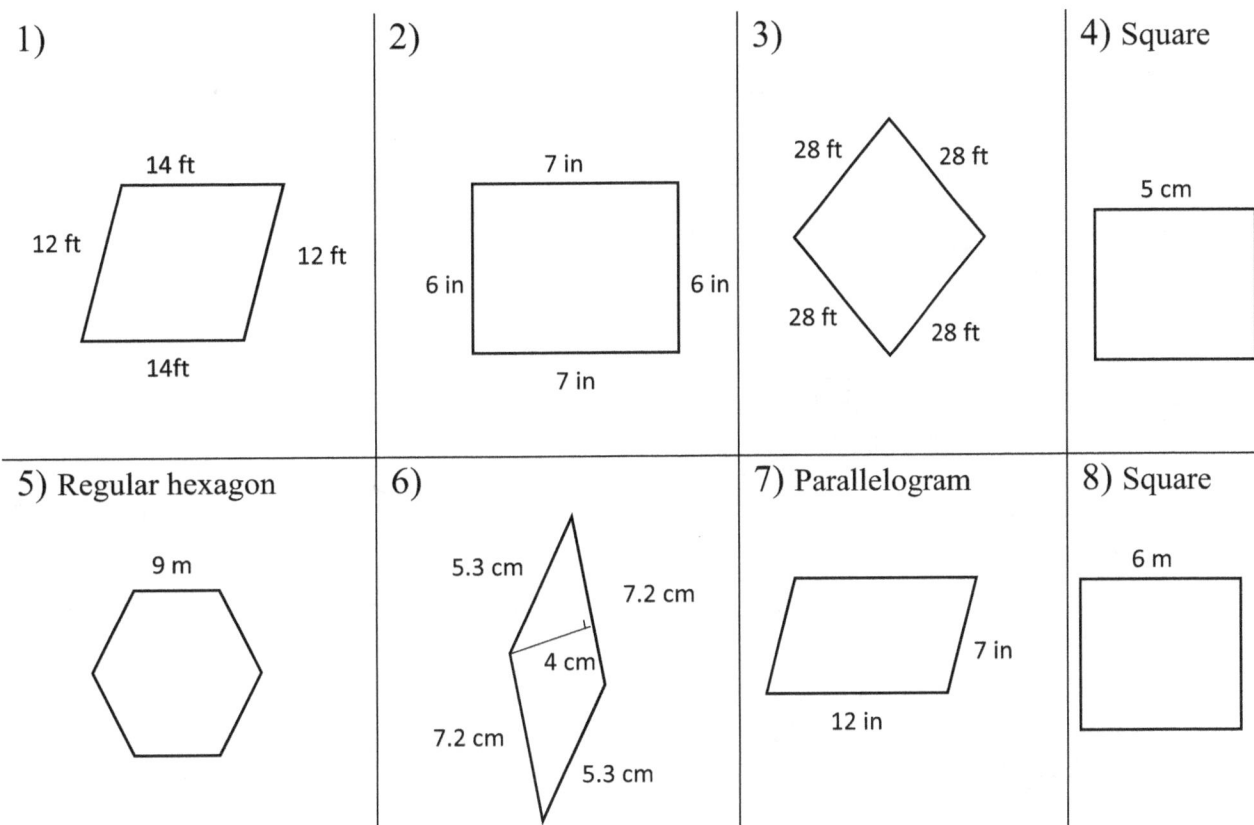

✍ **Find the area of each shape.**

9) Parallelogram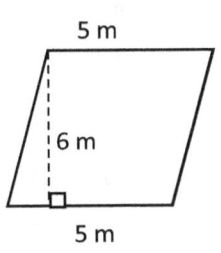

10) Rectangle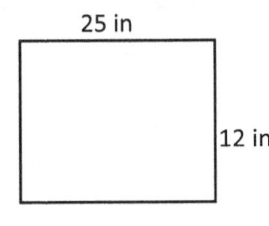

11) Rectangle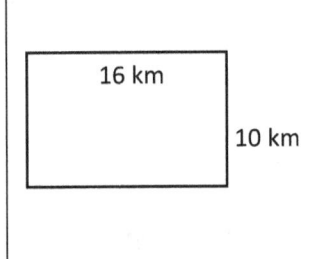

12) Square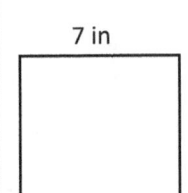

GED Subject Test – Mathematics

Trapezoids

✏️ **Find the area of each trapezoid.**

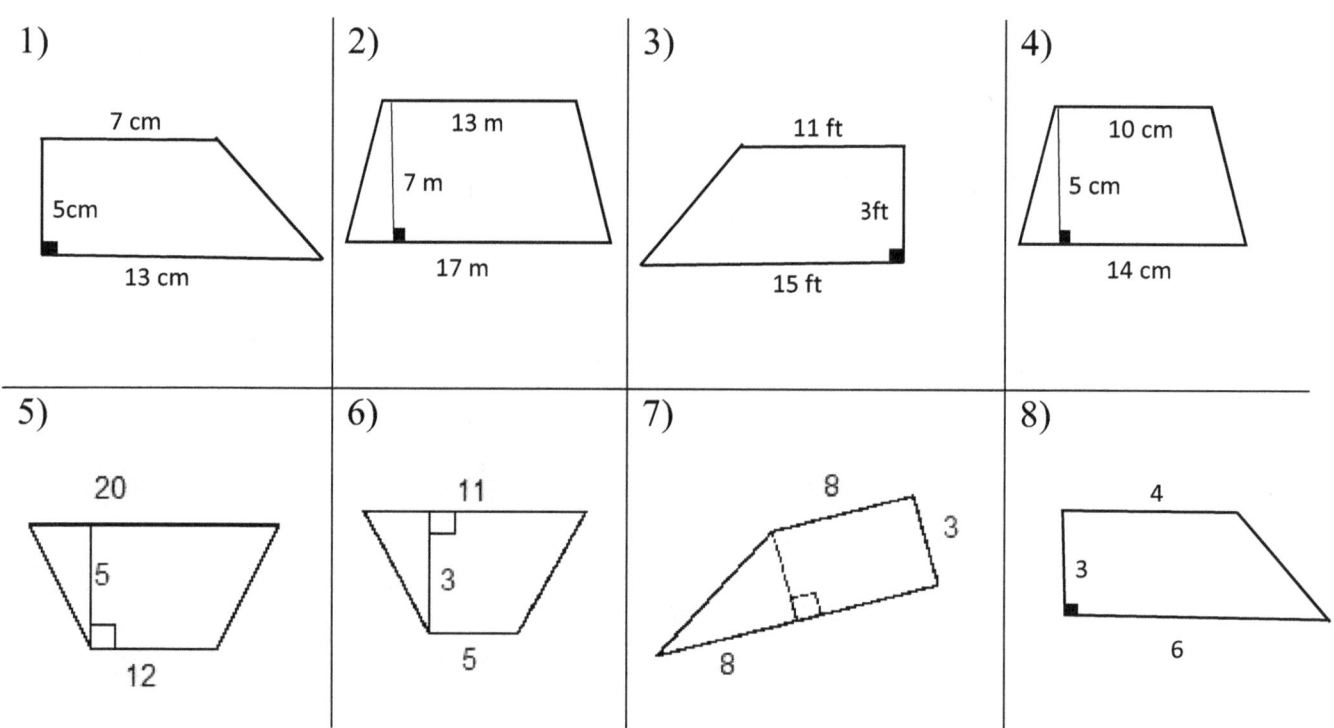

✏️ **Calculate.**

1) A trapezoid has an area of 45 cm² and its height is 5 cm and one base is 5 cm. What is the other base length? _____

2) If a trapezoid has an area of 99 ft² and the lengths of the bases are 8 ft and 10 ft, find the height? _____

3) If a trapezoid has an area of 126 m² and its height is 14 m and one base is 6 m, find the other base length? _____

4) The area of a trapezoid is 440 ft² and its height is 22 ft. If one base of the trapezoid is 15 ft, what is the other base length?

_____

Circles

Find the area of each circle. ($\pi = 3.14$)

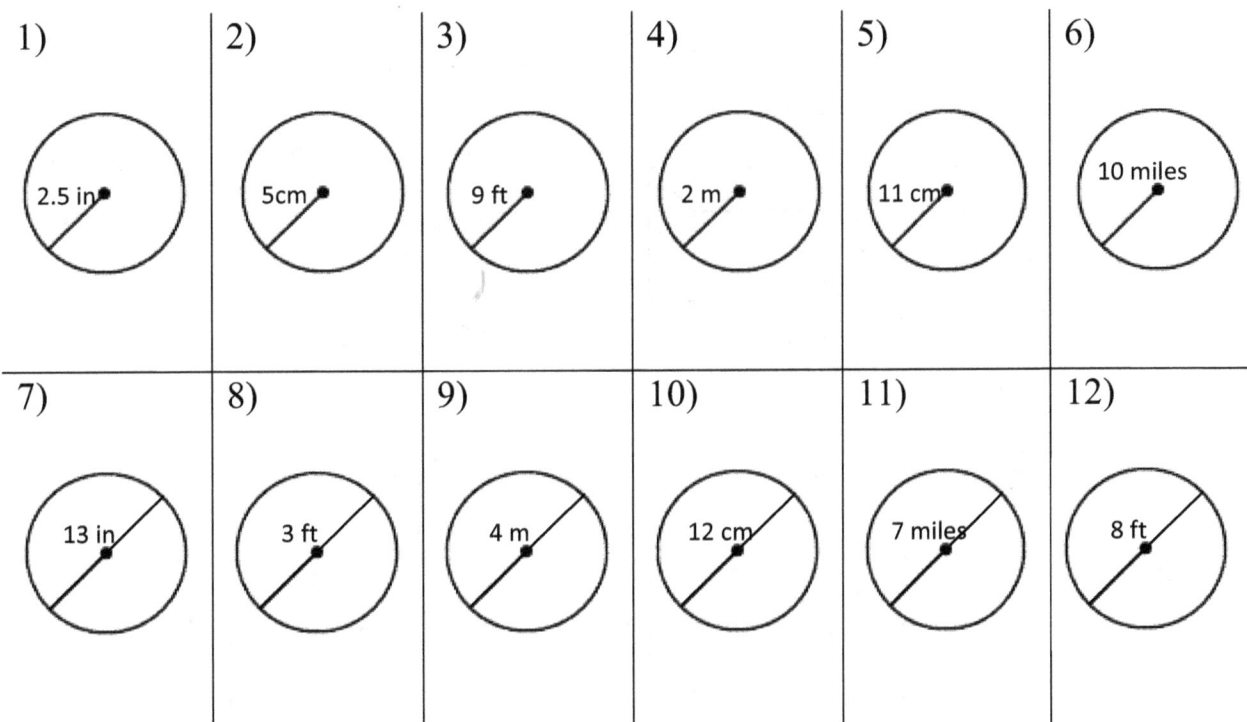

1) 2.5 in
2) 5 cm
3) 9 ft
4) 2 m
5) 11 cm
6) 10 miles
7) 13 in
8) 3 ft
9) 4 m
10) 12 cm
11) 7 miles
12) 8 ft

Complete the table below. ($\pi = 3.14$)

Circle No.	Radius	Diameter	Circumference	Area
1	1 in	2 in	6.28 in	3.14 in^2
2		10 m		
3				28.26 ft^2
4			47.1 mi	
5		11 km		
6	7 cm			
7		12 ft		
8				314 m^2
9			56.52 in	
10	4.5 ft			

GED Subject Test – Mathematics

Cubes

✏ **Find the volume of each cube.**

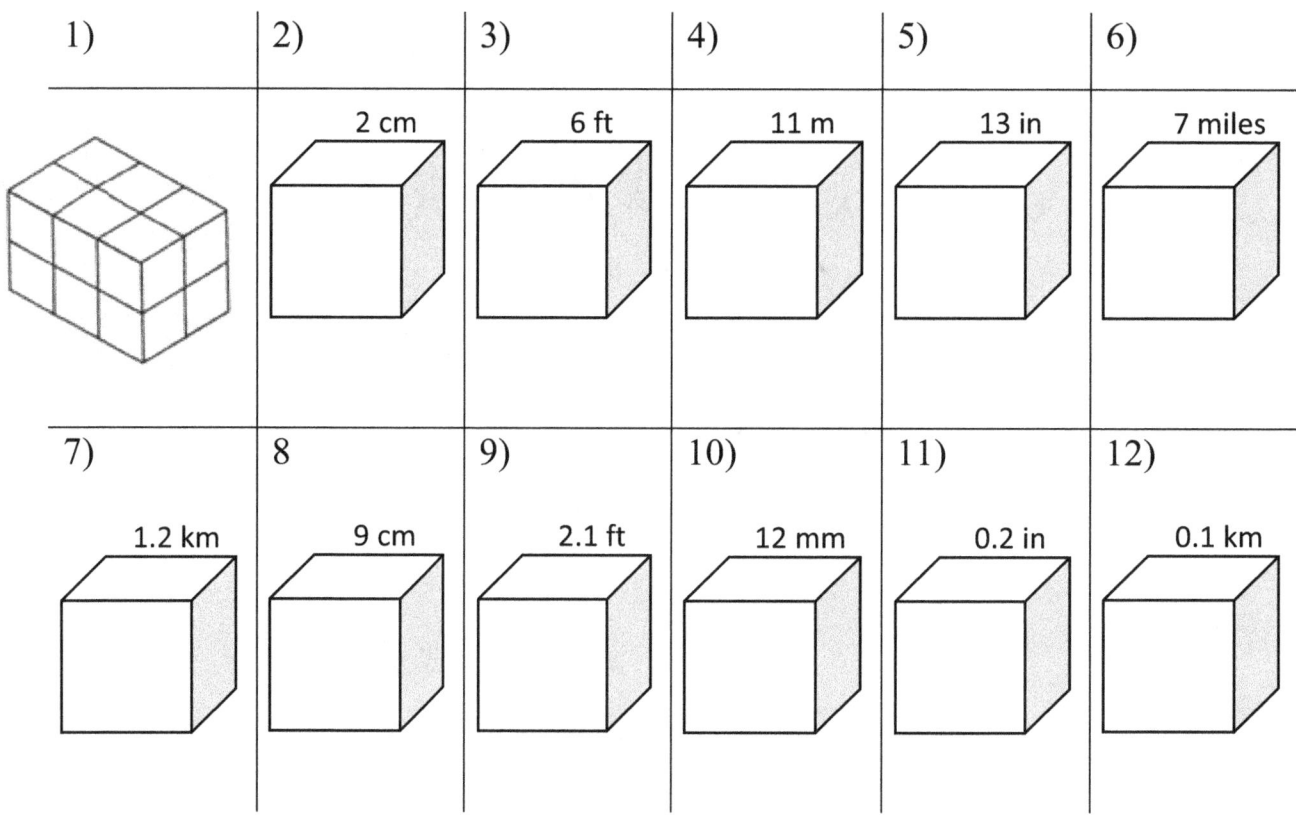

✏ **Find the surface area of each cube.**

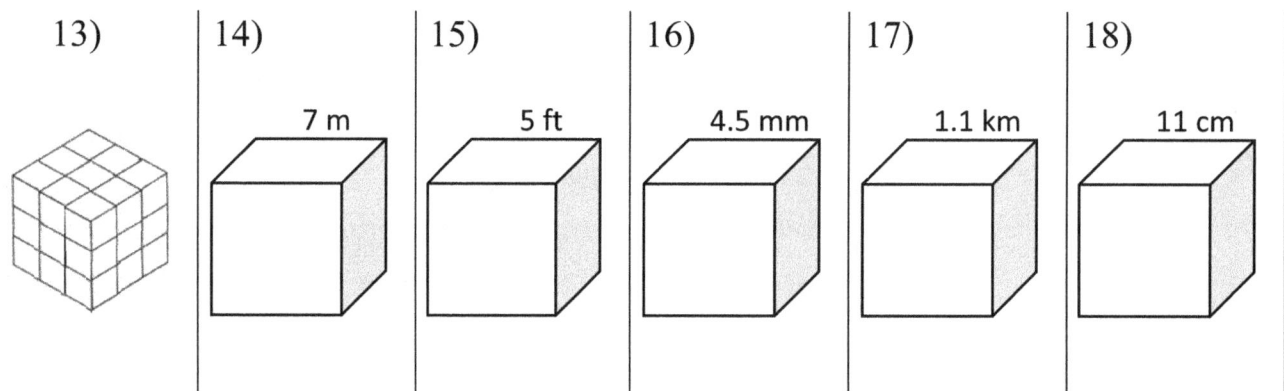

WWW.MathNotion.Com

GED Subject Test – Mathematics

Rectangular Prism

✍ **Find the volume of each Rectangular Prism.**

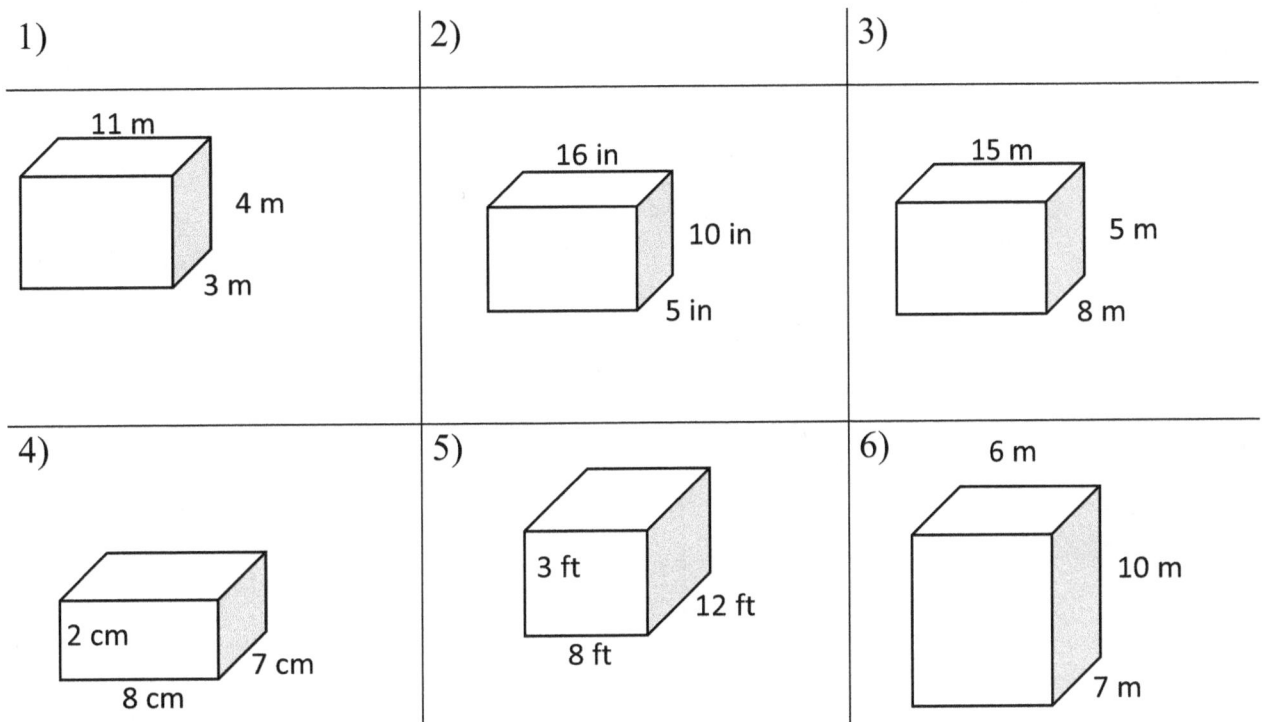

✍ **Find the surface area of each Rectangular Prism.**

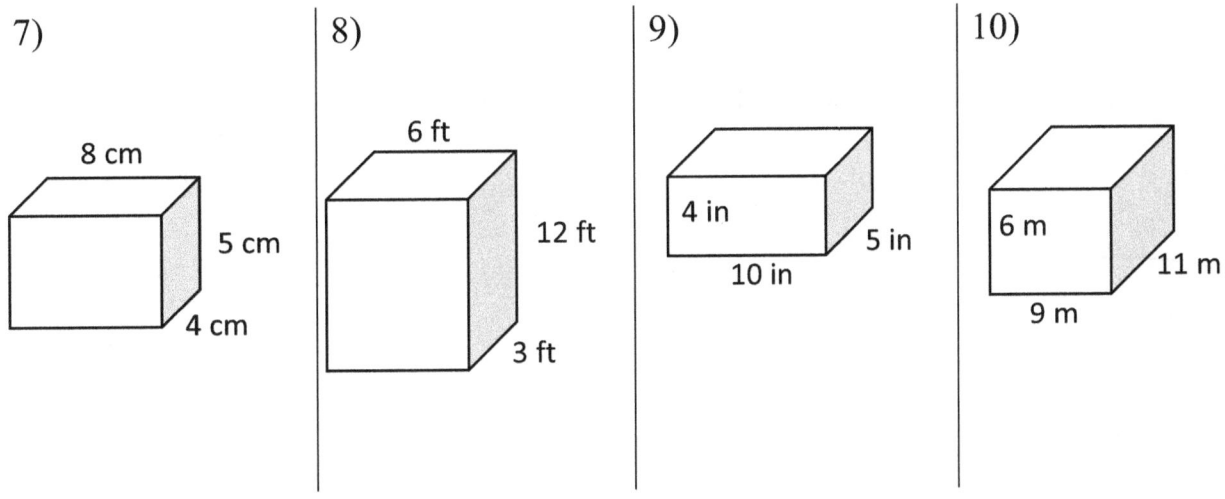

WWW.MathNotion.Com

GED Subject Test – Mathematics

Cylinder

✎ **Find the volume of each Cylinder. Round your answer to the nearest tenth.** ($\pi = 3.14$)

1)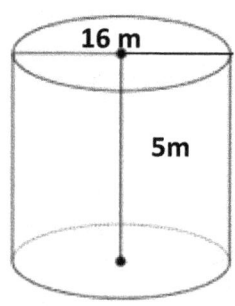

2)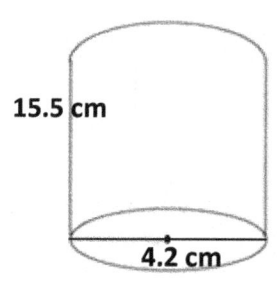

3)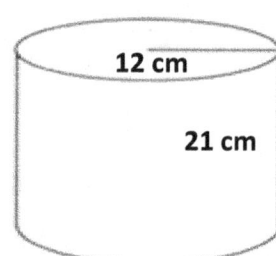

4)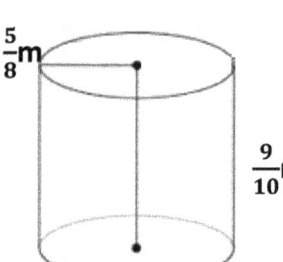

5)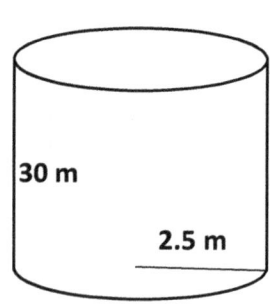

6)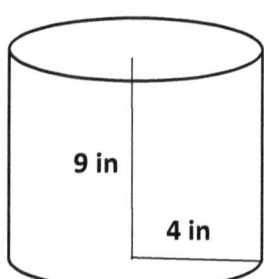

✎ **Find the surface area of each Cylinder.** ($\pi = 3.14$)

7)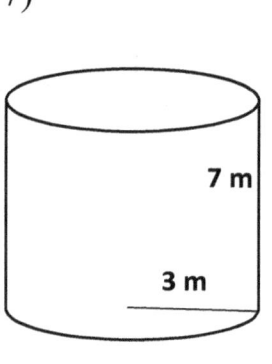

8)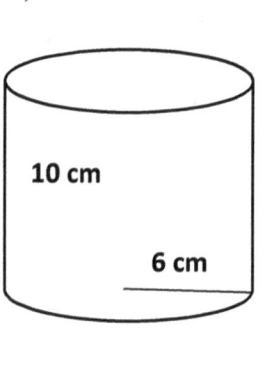

9)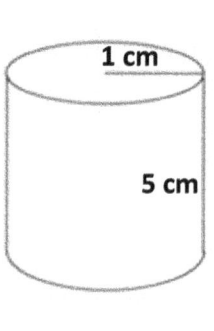

10)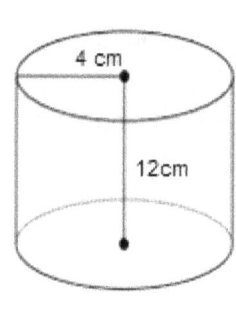

WWW.MathNotion.Com

GED Subject Test – Mathematics

Pyramids and Cone

✏️ **Find the volume of each Pyramid and Cone. ($\pi = 3.14$)**

1)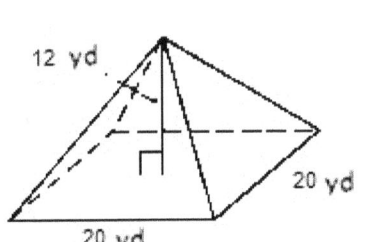

2)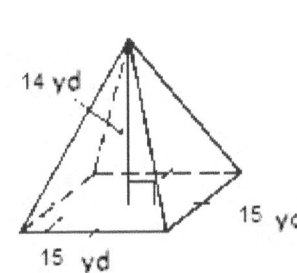

3)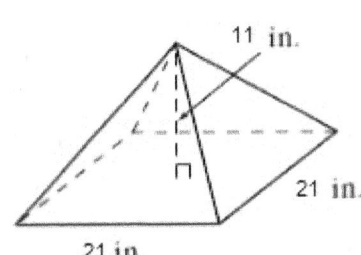

4)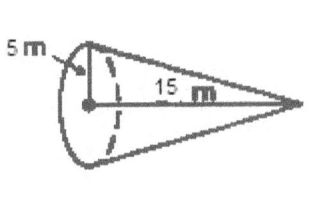

5)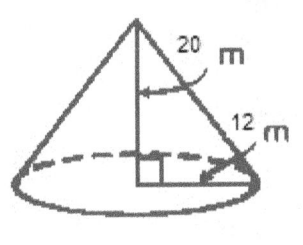

6)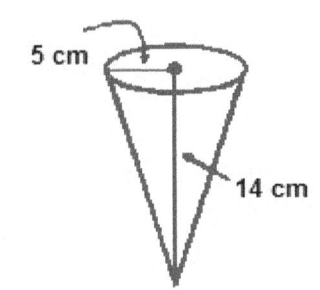

✏️ **Find the surface area of each Pyramid and Cone. ($\pi = 3.14$)**

7)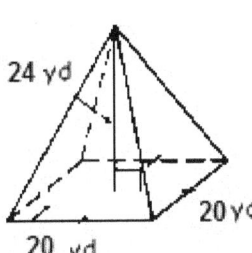

8)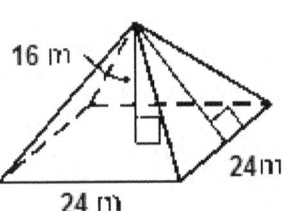

9)

10)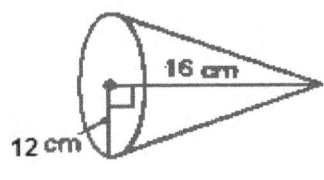

WWW.MathNotion.Com

GED Subject Test – Mathematics

Answers of Worksheets

Angles

1) 16° 4) 34° 7) 90° 10) 75°
2) 96° 5) 70° 8) 75° 11) 70°
3) 59° 6) 52° 9) 33°

Pythagorean Relationship

1) No   5) Yes 9) 13   13) 15
2) Yes 6) No   10) 20 14) 30
3) No   7) Yes 11) 17 15) 36
4) Yes 8) Yes 12) 10 16) 12

Triangles

1) 60° 5) 45° 9) 54 square unites
2) 48° 6) 40° 10) 120 square unites
3) 55° 7) 45° 11) 90 square unites
4) 52° 8) 67° 12) 180 square unites

Polygons

1) 52 ft 5) 54 m 9) 30 m^2
2) 26 in 6) 25 cm 10) 300 in^2
3) 112 ft 7) 38 in 11) 160 km^2
4) 20 cm 8) 24 m 12) 49 in^2

Trapezoids

1) 50 cm^2 4) 60 cm^2 7) 36
2) 105 m^2 5) 80   8) 15
3) 39 ft^2 6) 24

Calculate

1) 13 cm 2) 11 ft 3) 12 m 4) 25 ft

Circles

1) 19.63 in^2 5) 379.94 cm^2 9) 12.56 m^2
2) 78.5 cm^2 6) 314 $miles^2$ 10) 113.04 cm^2
3) 254.34 ft^2 7) 132.67 in^2 11) 38.47 $miles^2$
4) 12.56 m^2 8) 7.07 ft^2 12) 50.24 ft^2

WWW.MathNotion.Com

GED Subject Test – Mathematics

Circle No.	Radius	Diameter	Circumference	Area
1	1 in	2 in	6.28 in	3.14 in^2
2	5 m	10 m	31.4 m	78.5 m^2
3	3 ft	6 ft	18.84 ft	28.26 ft^2
4	7.5 miles	15 mi	47.1 mi	176.63 mi^2
5	5.5 km	11 km	34.54 km	94.99 km^2
6	7 cm	14 cm	43.96 cm	153.86 cm^2
7	6 ft	12 ft	37.68 feet	113.04 ft^2
8	10 m	20 m	62.8 m	314 m^2
9	9 in	18 in	56.52 in	254.34 in^2
10	4.5 ft	9 ft	28.26 ft	63.585 ft^2

Cubes

1) 12
2) 8 cm^3
3) 216 ft^3
4) 1,331 m^3
5) 2,197 in^3
6) 343 $miles^3$
7) 1.728 km^3
8) 729 cm^3
9) 9.261 ft^3
10) 1,728 mm^3
11) 0.008 in^3
12) 0.001 km^3
13) 27
14) 294 m^2
15) 150 ft^2
16) 121.5 mm^2
17) 7.26 km^2
18) 726 cm^2

Rectangular Prism

1) 132 m^3
2) 800 in^3
3) 600 m^3
4) 112 cm^3
5) 288 ft^3
6) 420 m^3
7) 184 cm^2
8) 252 ft^2
9) 220 in^2
10) 438 m^2

Cylinder

1) 1,004.8 m^3
2) 214.6 cm^3
3) 9,495.4 cm^3
4) 1.1 m^3
5) 588.8 m^3
6) 452.2 in^3
7) 188.4 m^2
8) 602.9 cm^2
9) 37.7 cm^2
10) 401.9 m^2

Pyramids and Cone

1) 1,600 yd^3
2) 1,050 yd^3
3) 1,617 in^3
4) 392.5 m^3
5) 3,014.4 m^3
6) 366.33 cm^3
7) 1,440 yd^2
8) 1,536 m^2
9) 678.24 in^2
10) 1,205.76 cm^2

GED Subject Test – Mathematics

GED Subject Test – Mathematics

Chapter 11: Statistics and Probability

Topics that you'll practice in this chapter:

- ✓ Mean and Median
- ✓ Mode and Range
- ✓ Histograms
- ✓ Stem–and–Leaf Plot
- ✓ Pie Graph
- ✓ Probability Problems

Mathematics is no more computation than typing is literature.

– John Allen Paulos

GED Subject Test – Mathematics

Mean and Median

✎ **Find Mean and Median of the Given Data.**

1) 8, 7, 14, 4, 8

2) 14, 8, 25, 19, 16, 33, 11

3) 23, 18, 15, 12, 17

4) 34, 14, 10, 15, 6, 11

5) 10, 19, 6, 8, 32, 20, 17

6) 17, 26, 39, 69, 20, 6

7) 40, 38, 18, 11, 9, 2, 7, 32, 41

8) 24, 21, 31, 12, 33, 32, 22

9) 16, 14, 20, 41, 15, 20, 38, 4

10) 20, 20, 30, 18, 6, 28, 12, 46

11) 12, 7, 10, 11, 16, 22

12) 10, 29, 27, 12, 2, 15, 10, 3

✎ **Calculate.**

13) In a javelin throw competition, five athletics score 56, 34, 62, 23 and 19 meters. What are their Mean and Median? _____

14) Eva went to shop and bought 8 apples, 14 peaches, 6 bananas, 4 pineapples and 12 melons. What are the Mean and Median of her purchase? _____

15) Bob has 17 black pen, 19 red pen, 14 green pens, 20 blue pens and 5 boxes of yellow pens. If the Mean and Median are 19 respectively, what is the number of yellow pens in each box? _____

WWW.MathNotion.Com

GED Subject Test – Mathematics

Mode and Range

✎ **Find Mode and Rage of the Given Data.**

1) 4, 3, 7, 3, 3, 4
   Mode: _____   Range: _____

2) 18, 18, 24, 26, 18, 8, 14, 22
   Mode: _____   Range: _____

3) 8, 8, 8, 16, 19, 22, 20, 9, 13
   Mode: _____   Range: _____

4) 24, 24, 14, 28, 20, 18, 20, 24
   Mode: _____   Range: _____

5) 6, 21, 27, 24, 27, 27
   Mode: _____   Range: _____

6) 21, 8, 8, 7, 8, 12, 10, 22, 18, 13
   Mode: _____   Range: _____

7) 7, 4, 4, 6, 13, 13, 13, 0, 2, 2
   Mode: _____   Range: _____

8) 5, 8, 5, 14, 12, 14, 3, 5, 18
   Mode: _____   Range: _____

9) 7, 7, 7, 12, 7, 3, 8, 16, 3, 17
   Mode: _____   Range: _____

10) 15, 15, 19, 16, 4, 16, 10, 15
    Mode: _____   Range: _____

11) 6, 6, 5, 6, 42, 13, 19, 2
    Mode: _____   Range: _____

12) 8, 8, 9, 8, 9, 4, 34, 22
    Mode: _____   Range: _____

✎ **Calculate.**

13) A stationery sold 12 pencils, 56 red pens, 24 blue pens, 20 notebooks, 12 erasers, 21 rulers and 11 color pencils. What are the Mode and Range for the stationery sells?

   Mode: _____   Range: _____

14) In an English test, eight students score 10, 15, 15, 18 18, 16, 15 and 15. What are their Mode and Range? _____

15) What is the range of the first 6 even numbers greater than 8?
    _____

WWW.MathNotion.Com

GED Subject Test – Mathematics

Times Series

👉 Use the following Graph to complete the table.

Day	Distance (km)
1	
2	

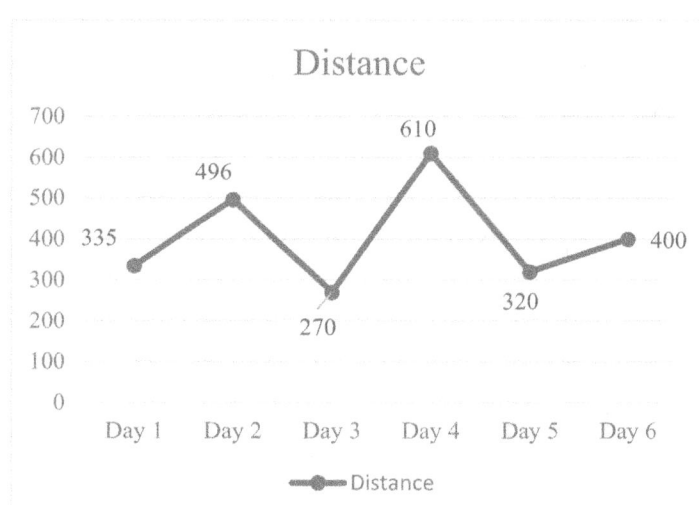

The following table shows the number of births in the US from 2007 to 2012 (in millions).

Year	Number of births (in millions)
2007	4.15
2008	3.70
2009	3.45
2010	3.20
2011	1.75
2012	2.98

Draw a Time Series for the table.

WWW.MathNotion.Com

GED Subject Test – Mathematics

Stem–and–Leaf Plot

✏️ **Make stem ad leaf plots for the given data.**

1) 24, 26, 29, 20, 53, 27, 51, 55, 36, 21, 37, 30

   Stem | Leaf plot

2) 11, 59, 66, 14, 18, 19, 59, 65, 69, 61, 68, 65

   Stem | Leaf plot

3) 121, 55, 66, 54, 112, 128, 63, 125, 59, 123, 68, 119

   Stem | Leaf plot

4) 51, 32, 100, 56, 84, 36, 107, 56, 85, 39, 56, 106, 89

   Stem | Leaf plot

5) 33, 89, 19, 87, 81, 16, 11, 30, 86, 35, 17, 35, 13

   Stem | Leaf plot

6) 60, 92, 22, 25, 67, 93, 95, 62, 21, 64, 98, 29

   Stem | Leaf plot

WWW.MathNotion.Com

GED Subject Test – Mathematics

Pie Graph

The circle graph below shows all Robert's expenses for last month. Robert spent $140 on his hobbies last month.

Answer following questions based on the Pie graph.

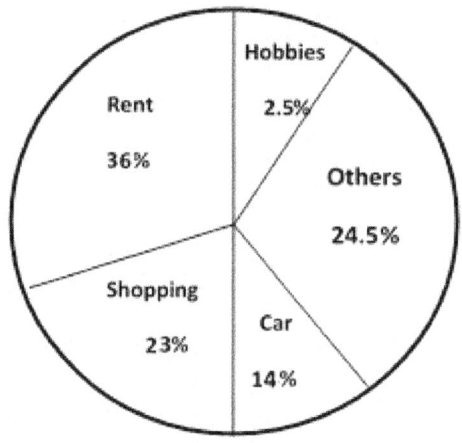

1) How much was Robert's total expenses last month? _____

2) How much did Robert spend on his car last month? _____

3) How much did Robert spend for shopping last month? _____

4) How much did Robert spend on his rent last month? _____

5) What fraction is Robert's expenses for his rent and car out of his total expenses last month? _____

GED Subject Test – Mathematics

Probability Problems

✎ **Calculate.**

1) A number is chosen at random from 1 to 10. Find the probability of selecting number 6 or smaller numbers. _____

2) Bag A contains 18 red marbles and 6 green marbles. Bag B contains 16 black marbles and 8 orange marbles. What is the probability of selecting a green marble at random from bag A? What is the probability of selecting a black marble at random from Bag B? _____

3) A number is chosen at random from 1 to 20. What is the probability of selecting multiples of 4? _____

4) A card is chosen from a well-shuffled deck of 52 cards. What is the probability that the card will be a queen? _____

5) A number is chosen at random from 1 to 15. What is the probability of selecting a multiple of 3 or 5? _____

A spinner numbered 1–8, is spun once. What is the probability of spinning …?

6) an Odd number? _____   7) a multiple of 2? _____

8) a multiple of 5? _____   9) number 10? _____

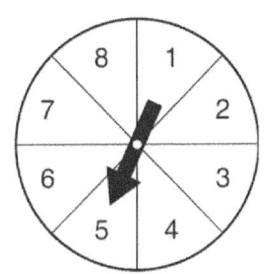

WWW.MathNotion.Com

GED Subject Test – Mathematics

Answers of Worksheets

Mean and Median

1) Mean: 8.2, Median: 8
2) Mean: 18, Median: 16
3) Mean: 17, Median: 17
4) Mean: 15, Median: 12.5
5) Mean: 16, Median: 17
6) Mean: 29.5, Median: 23
7) Mean: 22, Median: 18
8) Mean: 25, Median: 24
9) Mean: 21, Median: 18
10) Mean: 22.5, Median: 20
11) Mean: 13, Median: 11.5
12) Mean: 13.5, Median: 11
13) Mean: 38.8, Median: 34
14) Mean: 8.8, Median: 8
15) 5

Mode and Range

1) Mode: 3, Range: 4
2) Mode: 18, Range: 18
3) Mode: 8, Range: 14
4) Mode: 24, Range: 14
5) Mode: 27, Range: 21
6) Mode: 8, Range: 15
7) Mode: 13, Range: 13
8) Mode: 5, Range: 15
9) Mode: 7, Range: 14
10) Mode: 15, Range: 15
11) Mode: 6, Range: 40
12) Mode: 8, Range: 30
13) Mode: 12, Range: 45
14) Mode: 15, Range: 8
15) 10

Time series

Day	Distance (km)
1	335
2	496
3	270
4	610
5	320
6	400

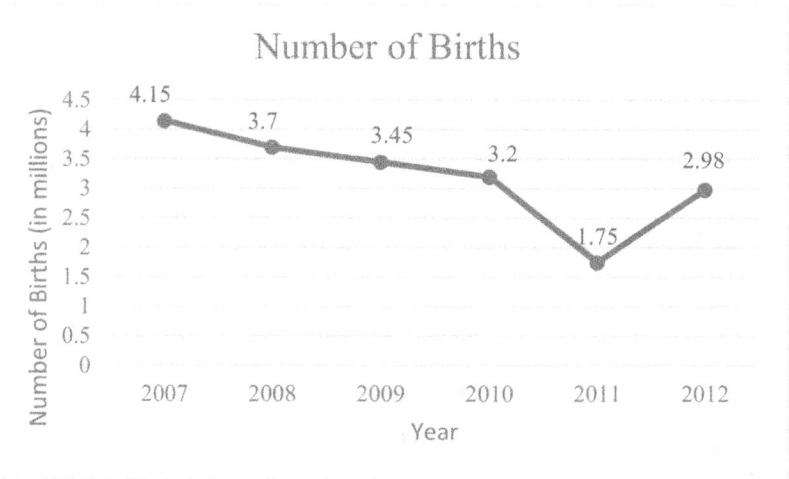

Stem–And–Leaf Plot

1)

Stem	leaf
2	0 1 4 6 7 9
3	0 6 7
5	1 3 5

2)

Stem	leaf
1	1 4 8 9
5	9 9
6	1 5 5 6 8 9

3)

Stem	leaf
5	4 5 9
6	3 6 8
11	2 9
12	1 3 5 8

WWW.MathNotion.Com

GED Subject Test – Mathematics

4)

Stem	leaf
3	2 6 9
5	1 6 6 6
8	4 5 9
10	0 6 7

5)

Stem	leaf
1	1 3 6 7 9
3	0 3 5 5
8	1 6 7 9

6)

Stem	leaf
2	2 1 5 9
6	0 2 4 7
9	2 3 5 8

Pie Graph

1) $5,600   3) $1,288   5) $\frac{1}{2}$

2) $784   4) $2,016

Probability Problems

1) $\frac{3}{5}$ 4) $\frac{1}{13}$ 7) $\frac{1}{2}$

2) $\frac{1}{4}, \frac{2}{3}$ 5) $\frac{7}{15}$ 8) $\frac{1}{8}$

3) $\frac{1}{4}$ 6) $\frac{1}{2}$ 9) 0

GED Subject Test – Mathematics

GED Subject Test – Mathematics

Chapter 12 : GED Test Review

The General Educational Development Test, commonly known as the GED or high school equivalency degree, is a standardized test and is the only high school equivalency test recognized in all 50 USA states.

Currently, GED is a computer-based test. Official computer-based tests are given at test centers all over the country.

There are four subject area tests on GED:

- Reasoning Through Language Arts,
- Mathematical Reasoning,
- Social Studies,
- Science

The GED Mathematical Reasoning test is a 115-minute, single-section test that covers basic mathematics topics, quantitative problem-solving and algebraic questions. There are two parts on Mathematical Reasoning section. The first part contains 5 questions where calculators are not permitted. The second part contains 41 test questions. Calculator is allowed in the second part.

In this section, there are two complete GED Mathematical Reasoning Tests. Take these tests to see what score you'll be able to receive on a real GED test.

GED Subject Test – Mathematics

Time to Test

Time to refine your skill with a practice examination

Take a REAL GED Mathematical Reasoning test to simulate the test day experience. After you've finished, score your test using the answer key.

Before You Start

- You'll need a pencil, calculator, and a timer to take the test.

- It's okay to guess. You won't lose any points if you're wrong.

- After you've finished the test, review the answer key to see where you went wrong.

Calculators are permitted only for the Part 2 of the test.

Good Luck!

GED Subject Test – Mathematics

GED Test Mathematics Formula Sheet

Area of a:

Parallelogram $\qquad A = bh$

Trapezoid $\qquad A = \frac{1}{2}h(b_1 + b_2)$

Surface Area and Volume of a:

Rectangular/Right Prism	$SA = ph + 2B$	$V = Bh$
Cylinder	$SA = 2\pi rh + 2\pi r^2$	$V = \pi r^2 h$
Pyramid	$SA = \frac{1}{2}ps + B$	$V = \frac{1}{3}Bh$
Cone	$SA = \pi rs + \pi r^2$	$V = \frac{1}{3}\pi r^2 h$
Sphere	$SA = 4\pi r^2$	$V = \frac{4}{3}\pi r^3$

(p = perimeter of base B; $\pi = 3.14$)

Algebra

Slope of a line $\qquad m = \dfrac{y_2 - y_1}{x_2 - x_1}$

Slope-intercept form of the equation of a line $\qquad y = mx + b$

Point-slope form of the Equation of a line $\qquad y - y_1 = m(x - x_1)$

Standard form of a Quadratic equation $\qquad y = ax^2 + bx + c$

Quadratic formula $\qquad x = \dfrac{-b \pm \sqrt{b^2 - 4ac}}{2a}$

Pythagorean theorem $\qquad a^2 + b^2 = c^2$

Simple interest $\qquad I = prt$

(I= interest, p= principal, r= rate, t= time)

GED Subject Test – Mathematics

GED Subject Test – Mathematics

GED Practice Test 1

Mathematical Reasoning

Two Parts

Total number of questions: 46

Part 1 (Non-Calculator): 5 questions

Part 2 (Calculator): 41 questions

Total time for two Part: 115 Minutes

Administered *Month Year*

GED Subject Test – Mathematics

GED Practice Test 1

Mathematical Reasoning

Part 1: Non-Calculator

5 Questions

You may NOT use a calculator on this part.

Administered *Month Year*

GED Subject Test – Mathematics

1) A tree 55 feet tall casts a shadow 44 feet long. Jack is 5 feet tall. How long is Jack's shadow?

   A. 4 ft

   B. 2.6 ft

   C. 3.75 ft

   D. 4.2 ft

2) What is the value of the expression $-2(x + 5y) + 4(2 - x)^2$ when $x = 1$ and $y = -3$?

   A. -32

   B. -24

   C. 32

   D. 24

3) What is the slope of a line that is perpendicular to the line $2x - 5y = 9$?

   A. -5

   B. $-\frac{5}{2}$

   C. $\frac{5}{2}$

   D. 5

4) $[7 \times (-13)] + ([(-6) \times (-12)] \div 9) + (-23) = ?$

   Write your answer in the box below.

   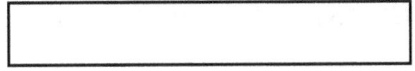

5) What is the product of all possible values of x in the following equation?

   $$|-4x + 2| = 14$$

   A. 3

   B. -12

   C. 12

   D. -4

GED Practice Test 1

Mathematical Reasoning

Part 2: Calculator

41 Questions

You may use a calculator on this part.

Administered

GED Subject Test – Mathematics

6) Simplify the expression.

$$(5x^5 - 3x^2 - 2x^3) - (2x^2 - 3x^5 - 4x^3)$$

A. $-(8x^5 + 2x^3 + 5x^2)$

B. $(8x^5 + 2x^3 - 5x^2)$

C. $6(x^5 + x^3 - x^2)$

D. $(-8x^5 - 2x^3 + 5x^2)$

7) What is the value of 2^{10}?

Write your answer in the box below.

8) Last week 18,000 fans attended a football match. This week Three times as many bought tickets, but one ninth of them cancelled their tickets. How many are attending this week?

A. 6,000

B. 54,000

C. 48,000

D. 60,000

9) In two successive years, the population of a town is increased by 25% and 20%. What percent of the population is increased after two years?

A. 60%

B. 37%

C. 50%

D. 13 %

GED Subject Test – Mathematics

10) Which of the following graphs represents the compound inequality $6 \leq 2x + 4 < 20$?

A.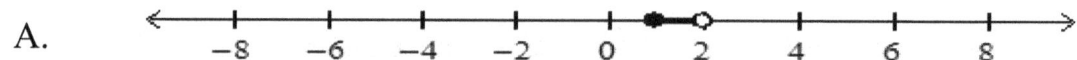

B.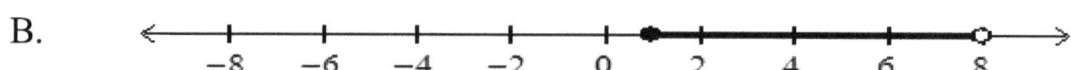

C.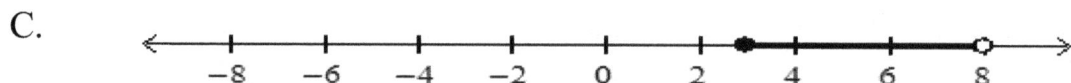

D.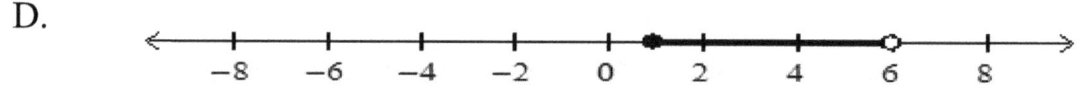

11) Two dice are thrown simultaneously, what is the probability of getting a sum of 5 or 7 or 10?

A. $\frac{11}{36}$

B. $\frac{13}{36}$

C. $\frac{7}{18}$

D. $\frac{4}{9}$

12) A swimming pool holds 6,144 cubic feet of water. The swimming pool is 32 feet long and 16 feet wide. How deep is the swimming pool?

Write your answer in the box below. (<u>Don't write the measurement</u>)

GED Subject Test – Mathematics

13) The mean of 35 test scores was calculated as 64. But it turned out that one of the scores was misread as 82 but it was 47. What is the correct mean of the test scores?

   A. 63

   B. 36

   C. 65.4

   D. 66.4

14) Which of the following shows the numbers in descending order?

$$\frac{1}{16}, 0.7, 25\%, \frac{3}{4}$$

   A. $\frac{1}{16}, 0.7, \frac{3}{4}, 25\%$

   B. $25\%, \frac{1}{16}, \frac{3}{4}, 0.7$

   C. $\frac{3}{4}, 0.7, 25\%, \frac{1}{16}$

   D. $\frac{1}{16}, 25\%, 0.7, \frac{3}{4}$

15) What is the volume of a box with the following dimensions?

   Height = 5 cm   Width = 7 cm   Length = 10 cm

   A. 175 cm³

   B. 350 cm³

   C. 70 cm³

   D. 22 cm³

16) The average of 9 numbers is 28. The average of 6 of those numbers is 35. What is the average of the other three numbers?

   A. 12

   B. 14

   C. 20

   D. 16

GED Subject Test – Mathematics

17) What is the value of y in the following system of equations?

$$2x + y = -2$$
$$-x - y = -2$$

A. -4

B. 6

C. 4

D. -6

18) The perimeter of a rectangular yard is 84 meters. What is its length if its width is twice its length?

A. 28 meters

B. 42 meters

C. 14 meters

D. 16 meters

19) In a stadium the ratio of home fans to visiting fans in a crowd is 5:7. Which of the following could be the total number of fans in the stadium? (Select one or more answer choices)

A. 42,000

B. 36,770

C. 52,300

D. 61,750

E. 33,000

20) What is the perimeter of a square in centimeters that has an area of 380.25 cm^2?

Write your answer in the box below. (don't write the measurement)

GED Subject Test – Mathematics

21) Anita's trick–or–treat bag contains 24 pieces of chocolate, 36 suckers, 20 pieces of gum, 30 pieces of licorice. If she randomly pulls a piece of candy from her bag, what is the probability of her pulling out a piece of gum?

A. $\frac{18}{55}$

B. $\frac{3}{11}$

C. $\frac{2}{11}$

D. $\frac{8}{55}$

22) What is the area of a square whose diagonal is 14?

A. 56

B. 196

C. 121

D. 98

23) Which of the following points lies on the line $6x - 2y = -10$? (Select one or more answer choices)

A. $(-3, -3)$

B. $(1, 8)$

C. $(-1, -2)$

D. $(0, 5)$

24) Mr. Carlos family are choosing a menu for their reception. They have 3 choices of appetizers, 5 choices of entrees, 10 choices of cake. How many different menu combinations are possible for them to choose?

A. 150

B. 30

C. 50

D. 300

GED Subject Test – Mathematics

25) The ratio of boys and girls in a class is 4:5. If there are 72 students in the class, how many more boys should be enrolled to make the ratio 1:1?

   A. 32

   B. 40

   C. 8

   D. 16

26) The average of three numbers is 64. If a fourth number that is greater than 78 is added, then, which of the following could be the new average?

   A. 62

   B. 56

   C. 68

   D. 65

27) The perimeter of the trapezoid below is 56 cm. What is its area?

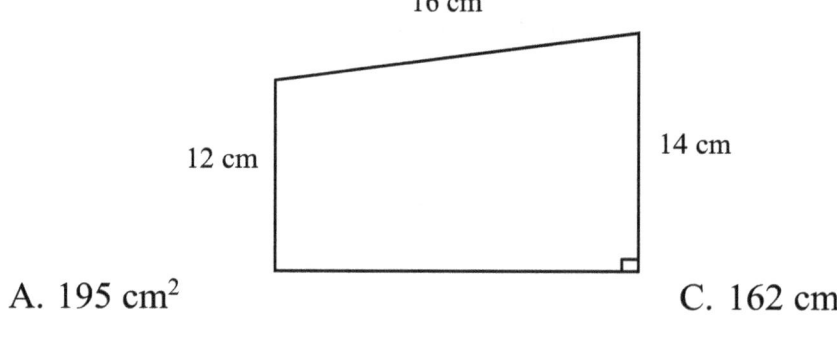

   A. 195 cm²

   B. 132 cm²

   C. 162 cm²

   D. 182 cm²

28) A card is drawn at random from a standard 52–card deck, what is the probability that the card is of Hearts or Diamonds? (The deck includes 13 of each suit clubs, diamonds, hearts, and spades)

   A. $\frac{1}{26}$

   B. $\frac{1}{4}$

   C. $\frac{1}{2}$

   D. $\frac{3}{13}$

GED Subject Test – Mathematics

29) The diagonal of a rectangle is 13 inches long and the height of the rectangle is 5 inches. What is the perimeter of the rectangle in inches?

Write your answer in the box below.

30) What is the surface area of the cylinder below?

   A. 90π in^2

   B. 282.6π in^2

   C. 216π in^2

   D. 326π in^2

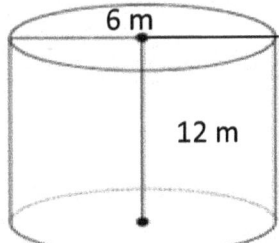

31) Simplify $4x^3y^6(-3xy^2)^2 =$

   A. $6x^4y^{12}$ C. $-12x^4y^{12}$

   B. $12x^5y^{10}$ D. $36x^5y^{10}$

32) Mr. Matthews saves $3,500 out of his annually family income of $84,000. What fractional part of his income does he save?

   A. $\frac{5}{12}$ C. $\frac{4}{21}$

   B. $\frac{1}{24}$ D. $\frac{5}{24}$

GED Subject Test – Mathematics

33) What is the median of these numbers? 29, 23, 18, 16, 19, 11, 10

    A. 23   C. 18

    B. 19.5   D. 16

34) A bank is offering 3.5% simple interest on a savings account. If you deposit $33,000, how much interest will you earn in five years?

    A. $7,750   C. $5,775

    B. $7,250   D. $38,775

35) Mrs. Thomson needs an 90% average in her writing class to pass. On her first 4 exams, he earned scores of 78%, 92%, 95%, and 96%. What is the minimum score Mrs. Thomson can earn on her fifth and final test to pass?

Write your answer in the box below.

☐

36) Daniel is 35 miles ahead of Noa and running at 4 miles per hour. Noa is running at the speed of 9 miles per hour. How long does it take Noa to catch Daniel?

    A. 6 hours, and 40 minutes   C. 8 hours

    B. 7 hours, 30 minutes   D. 7 hours

GED Subject Test – Mathematics

37) What is the equivalent temperature of $77°F$ in Celsius?

$$C = \frac{5}{9}(F - 32)$$

A. 22.8

B. 25

C. 45.5

D. 34

38) A football team had $52,000 to spend on supplies. The team spent $20,500 on new balls. New sport shoes cost $92 each. Which of the following inequalities represent the number of new shoes the team can purchase?

A. $92x + 20,500 \leq 52,000$

B. $92x + 20,500 \leq 52,000$

C. $20,500 + 92x \geq 52,000$

D. $20,500 + 92x \geq 52,000$

39) If 60% of a number is 45, what is the number?

A. 75

B. 65

C. 37.5

D. 60

40) The circle graph below shows all Mr. Wilson's expenses for last month. If he spent $840 on his car, how much did he spend for his rent?

A. $504

B. $1,260

C. $882

D. $714

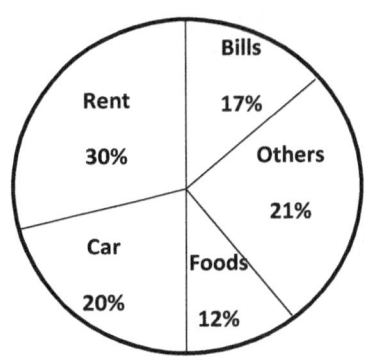

Mr. Green's monthly expenses

GED Subject Test – Mathematics

41) 165 students took an exam and 66 of them failed. What percent of the students passed the exam?

   A. 42 %   C. 66 %

   B. 40 %   D. 60 %

42) A cruise line ship left Port A and traveled 30 miles due west and then 40 miles due north. At this point, what is the shortest distance from the cruise to port A in miles?

Write your answer in the box below.

43) The square of a number is $\frac{75}{192}$. What is the cube of that number?

   A. $\frac{15}{64}$ C. $\frac{25}{81}$

   B. $\frac{125}{343}$ D. $\frac{125}{512}$

44) What is the value of x in the following equation?

$$\frac{1}{5}x + \frac{4}{15} = \frac{1}{2}$$

   A. 7   C. $\frac{7}{6}$

   B. $\frac{5}{6}$ D. $\frac{4}{15}$

GED Subject Test – Mathematics

45) If 125 % of a number is 85, then what is the 45% of that number?

   A. 30.60

   B. 27.20

   C. 61.20

   D. 36.25

46) What is the slope of the line?

$$-5x - 8y = 20$$

Write your answer in the box below.

☐

End of GED Mathematical Reasoning Practice Test 1

GED Subject Test – Mathematics

GED Subject Test – Mathematics

GED Practice Test 2

Mathematical Reasoning

Two Parts

Total number of questions: 46

Part 1 (Non-Calculator): 5 questions

Part 2 (Calculator): 41 questions

Total time for two Part: 115 Minutes

Administered *Month Year*

GED Subject Test – Mathematics

GED Practice Test 2

Mathematical Reasoning

Part 1: Non-Calculator

5 Questions

You may NOT use a calculator on this part.

Administered

GED Subject Test – Mathematics

1) Which of the following points lies on the line with equation $x - \frac{7}{5}y = -3$?

   A. $(-5, -10)$ C. $(-2, 5)$

   B. $(4, 5)$ D. $(4, -5)$

2) A shirt costing $530 is discounted 22%. After a month, the shirt is discounted another 13%. Which of the following expressions can be used to find the selling price of the shirt?

   A. $(530)(0.78)(0.87)$ C. $(530)(0.78) - (530)(0.13)$

   B. $(530)(0.22)(0.13)$ D. $(530)(0.22) - (530)(0.13)$

3) If $-2x + 5 = 9$, What is the value of $-2x - \frac{3}{4}$?

   A. 4.75   C. -4.75

   B. -2 D. 3.25

4) $11 \times (-2) + 8 + 2(-15 - 7 \times 3) \div 12 = ?$

   Write your answer in the box below.

   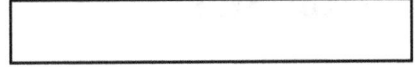

5) What is the area of an isosceles right triangle that has one leg that measures 6cm?

   A. 36 cm²   C. 20 cm²

   B. 18 cm²   D. 12 cm²

GED Subject Test – Mathematics

GED Practice Test 2

Mathematical Reasoning

Part 2: Calculator

41 Questions

You may use a calculator on this part.

Administered

GED Subject Test – Mathematics

6) Ryan traveled 450 km in 9 hours and Riley traveled 490 km in 7 hours. What is the ratio of the average speed of Ryan to average speed of Riley?

   A. 6 : 5

   B. 11 : 2

   C. 5 : 7

   D. 3 : 4

7) What is the value of x in the following system of equation?

$$4x + 3y = 5$$

$$-x - 2y = -5$$

Write your answer in the box below.

☐

8) An angle is equal to one ninth of its supplement. What is the measure of that angle?

   A. 29

   B. 24

   C. 18

   D. 54

9) Abigail purchased a sofa for $369. The sofa is regularly priced at $450. What was the percent discount Abigail received on the sofa?

   A. 18%

   B. 12.20 %

   C. 36%

   D. 11.80 %

GED Subject Test – Mathematics

10) When a number is subtracted from 90 and the difference is divided by that number, the result is 5. What is the value of the number?

   A. 15   C. 4

   B. 5   D. 10

11) Right triangle ABC has two legs of lengths 5 cm (AB) and 12 cm (AC). What is the length of the third side (BC)?

   A. 13 cm   C. 26 cm

   B. 15 cm   D. 16 cm

12) A taxi driver earns $18 per hour work. If he works 10 hours a day, and he uses 3-liters Petrol in 2 hours with price $3.25 for 1-liter. How much money does he earn in one day?

   A. $180   C. $132.75

   B. $48.75   D. $131.25

13) The width of a box is one fourth of its length. The height of the box is one tenth of its width. If the length of the box is 80 cm, what is the volume of the box?

   A. 1,280 cm^3 C. 3,200 cm^3

   B. 1,240 cm^3 D. 3,400 c m^3

WWW.MathNotion.Com

GED Subject Test – Mathematics

14) The price of a sofa is decreased by 10% to $1,350. What was its original price?

   A. $1,800

   B. $1,600

   C. $1,400

   D. $1,500

15) If 60% of a class are girls, and 40% of girls play tennis, what percent of the class play tennis?

   A. 20%

   B. 24%

   C. 68%

   D. 36%

16) The average of 42, 46, 40 and x is 44. What is the value of x?

   Write your answer in the box below.

   [ ]

17) The price of a car was $36,000 in 2014, $25,200 in 2015 and $17,640 in 2016. What is the rate of depreciation of the price of car per year?

   A. 30 %

   B. 40%

   C. 20%

   D. 15%

18) The average of seven consecutive numbers is 57. What is the smallest number?

   A. 53

   B. 52

   C. 54

   D. 55

WWW.MathNotion.Com

GED Subject Test – Mathematics

19) The area of a circle is less than 64π. Which of the following can be the circumference of the circle? (Select one or more answer choices)

A. 14π

B. 20π

C. 25π

D. 36π

E. 8π

20) A bank is offering 2.25% simple interest on a savings account. If you deposit $12,800, how much interest will you earn in six years?

A. $1,728

B. $2,827

C. $288

D. $1,440

21) In four successive hours, a car travels 56 km, 42 km, 68 km and 24km. In the next six hours, it travels with an average speed of 37 km per hour. Find the total distance the car traveled in 10 hours.

A. 480 km

B. 412 km

C. 580 km

D. 540 km

22) The ratio of boys to girls in a school is 2:7. If there are 288 students in a school, how many boys are in the school.

Write your answer in the box below.

GED Subject Test – Mathematics

23) How long does a 120–miles trip take moving at 16 miles per hour (mph)?

   A. 7 hours and 50 minutes

   B. 8 hours

   C. 7 hours and 30 minutes

   D. 8 hours and 30 minutes

24) The perimeter of the trapezoid below is 86. What is its area?

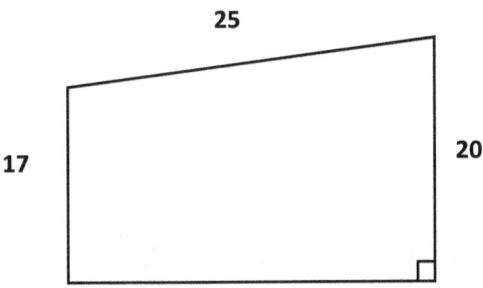

Write your answer in the box below.

25) In the xy-plane, the point $(1, -3)$ and $(4, 6)$ are on the line A. Which of the following points could also be on the line A? (Select one or more answer choices)

   A. $(5, 4)$

   B. $(-1, 7)$

   C. $(2, 0)$

   D. $(-3, 4)$

   E. $(-2, -12)$

26) 40 is What percent of 32?

   A. 25 %

   B. 125 %

   C. 15 %

   D. 115 %

GED Subject Test – Mathematics

27) The marked price of a computer is E Euro. Its price decreased by 20% in March and later increased by 5% in April. What is the final price of the computer in E Euro?

   A. 1.84 E   C. 0.85 E

   B. 0.84 E   D. 1.85 E

28) A rope weighs 340 grams per meter of length. What is the weight in kilograms of 25 meters of this rope? (1 kilograms = 1,000 grams)

   A. 8,500   C. 8.500

   B. 85.00   D. 0.8500

29) Which of the following could be the product of two consecutive prime numbers? (Select one or more answer choices)

   A. 21   D. 56

   B. 144   E. 154

   C. 221   F. 420

30) A $120 shirt now selling for $69 is discounted by what percent?

   A. 55.2 %   C. 42.5 %

   B. 22.4 %   D. 24 %

GED Subject Test – Mathematics

31) The score of Zoe was one fifth of Emma and the score of Harper was triple that of Emma. If the score of Harper was 189, what is the score of Zoe?

   A. 126

   B. 12.6

   C. 63

   D. 56

32) Three seventh of 63 is equal to $\frac{9}{11}$ of what number?

   A. 45

   B. 33

   C. 186

   D. 81

33) A bag contains 20 balls: four green, two black, five blue, seven red and two white. If 15 balls are removed from the bag at random, what is the probability that a blue ball has been removed?

   A. $\frac{1}{4}$

   B. $\frac{7}{20}$

   C. $\frac{2}{9}$

   D. $\frac{1}{20}$

34) What is the value of 9^3?

   Write your answer in the box below.

   [ ]

35) What is the median of these numbers? 7, 20, 20, 19, 14, 24, 11, 27, 26

   A. 20

   B. 26

   C. 14

   D. 24

WWW.MathNotion.Com

GED Subject Test – Mathematics

36) How many tiles of 8 cm² is needed to cover a floor of dimension 5 cm by 56 cm?

   A. 35.5   C. 35

   B. 32.5   D. 32

37) A chemical solution contains 6% alcohol. If there is 36.6 ml of alcohol, what is the volume of the solution?

   A. 410 ml   C. 6.1 ml

   B. 610 ml   D. 660 ml

38) The radius of the following cylinder is 6 inches, and its height are 10 inches. What is the surface area of the cylinder in square inches? (π equals 3.14)

   Write your answer in the box below.

   ☐

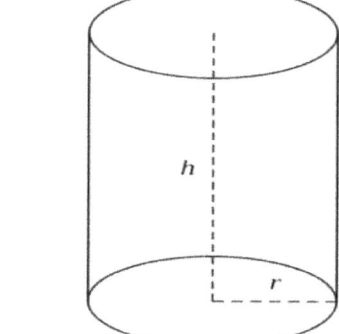

39) The average high of 24 constructions in a town is 135 m, and the average high of 16 towers in the same town is 165 m. What is the average high of all the 40 structures in that town?

   A. 147   C. 143.5

   B. 157   D. 137.2

GED Subject Test – Mathematics

40) The price of a laptop is decreased by 40% to $615. What is its original price?

   A. 1,250

   B. 550

   C. 1,537

   D. 125

41) In 1999, the average worker's income increased $3,420 per year starting from $48,000 annual salary. Which equation represents income greater than average? (I = income, x = number of years after 1999)

   A. $I > -3{,}420\,x + 48{,}000$

   B. $I > 3{,}420\,x + 48{,}000$

   C. $I < 48{,}000\,x + 3{,}420$

   D. $I < -48{,}000\,x + 3{,}420$

42) A boat sails 12 miles south and then 9 miles east. How far is the boat from its start point?

   A. 21 miles

   B. 108 miles

   C. 15 miles

   D. 30 miles

43) If 96% of F is 16% of M, then F is what percent of M?

   A. 600 %

   B. 1,400 %

   C. 60 %

   D. 300 %

GED Subject Test – Mathematics

44) Which graph corresponds to the following inequalities?

$$y \leq x + 3$$

$$x + 3y \geq 6$$

A.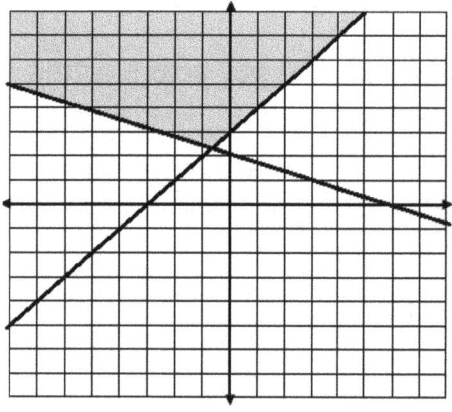

B.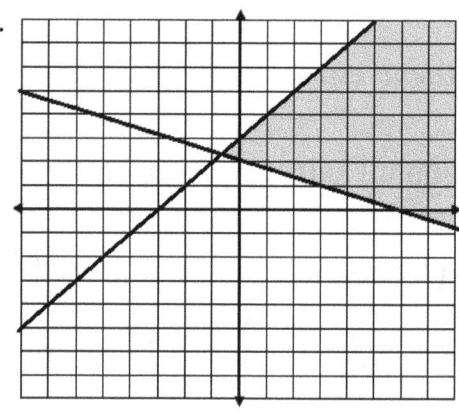

C.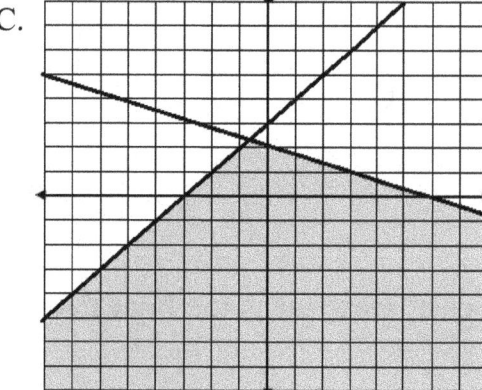

D.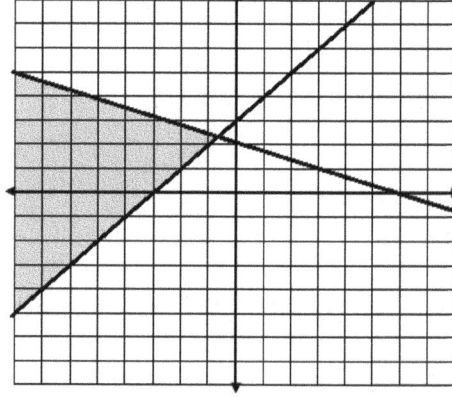

45) From last year, the price of gasoline has increased from $1.2 per gallon to $4.8 per gallon. The new price is what percent of the original price?

A. 400 %

B. 250 %

C. 480 %

D. 600 %

GED Subject Test – Mathematics

46) How many possible outfit combinations come from six shirts, four slacks, and 9 times?

Write your answer in the box below.

☐

End of GED Mathematical Reasoning Practice Test 2

GED Subject Test – Mathematics

GED Subject Test – Mathematics

Chapter 13 : Answers and Explanations

Answer Key

❋Now, it's time to review your results to see where you went wrong and what areas you need to improve!

GED Math Practice Test

Practice Test 1

1	A	16	B	31	D
2	C	17	B	32	B
3	B	18	C	33	C
4	−106	19	A-E	34	C
5	B	20	78	35	89
6	B	21	C	36	D
7	1,024	22	D	37	B
8	C	23	B–D	38	A
9	C	24	A	39	A
10	B	25	C	40	B
11	B	26	C	41	D
12	12	27	C	42	50
13	A	28	C	43	D
14	C	29	34	44	C
15	B	30	A	45	A
				46	$-\dfrac{5}{8}$

Practice Test 2

1	B	16	48	31	B
2	A	17	A	32	B
3	D	18	C	33	A
4	−20	19	A-E	34	729
5	B	20	A	35	A
6	C	21	B	36	C
7	−1	22	64	37	B
8	C	23	C	38	602.88
9	A	24	444	39	A
10	A	25	C-E	40	A
11	A	26	B	41	B
12	D	27	B	42	C
13	C	28	C	43	A
14	D	29	A-C	44	B
15	B	30	C	45	A
				46	216

GED Subject Test – Mathematics

GED Subject Test – Mathematics

Practice Test 1
GED Mathematical Reasoning

1) Answer: A

Write a proportion and solve for the missing number.

$\frac{55}{44} = \frac{5}{x} \to 55x = 5 \times 44 = 220 \to 55x = 220 \to x = \frac{220}{55} = 4$

2) Answer: C

Plug in the value of x and y. $-2(x + 5y) + 4(2 - x)^2$ when $x = 1$ and $y = -3$

$= -2(1 + 5(-3)) + 4(2 - 1)^2 = -2(1 - 15) + 4(2 - 1)^2 = (-2)(-14) + 4(1)^2$

$= 28 + 4 = 32$

3) Answer: B

The equation of a line in slope intercept form is: $y = mx + b$

Solve for y. $2x - 5y = 9 \to -5y = -2x + 9$

Divide both sides by (-5). Then: $y = \frac{2}{5}x - \frac{9}{5}$; The slope of this line is $\frac{2}{5}$.

The product of the slopes of two perpendicular lines is -1. Therefore, the slope of a line that is perpendicular to this line is:

$m_1 \times m_2 = -1 \Rightarrow \frac{2}{5} \times m_2 = -1 \Rightarrow m_2 = \frac{-1}{\frac{2}{5}} = -\frac{5}{2}$

4) Answer: -106

Use PEMDAS (order of operation):

$[7 \times (-13)] + ([(-6) \times (-12)] \div 9) + (-23) = [-91] + (72 \div 9) - 23 = -91 + 8 - 23 = -106$.

5) Answer: B

To solve absolute values equations, write two equations.

$-4x + 2$ can equal positive 14, or negative 14. Therefore,

$-4x + 2 = 14 \Rightarrow -4x = 12 \Rightarrow x = -3$

$-4x + 2 = -14 \Rightarrow -4x = -14 - 2 = -16 \Rightarrow x = 4$

Find the product of solutions: $-3 \times 4 = -12$

GED Subject Test – Mathematics

6) Answer: B

Simplify and combine like terms.

$(5x^5 - 3x^2 - 2x^3) - (2x^2 - 3x^5 - 4x^3) \Rightarrow (5x^5 - 3x^2 - 2x^3) - 2x^2 + 3x^5 + 4x^3 \Rightarrow 8x^5 + 2x^3 - 5x^2$.

7) Answer: 1,024

$2^{10} = 2 \times 2 \times 2 \times 2 \times 2 \times 2 \times 2 \times 2 \times 2 \times 2 = 1,024$

8) Answer: C

Three times of 18,000 is 54,000. One ninth of them cancelled their tickets.

One ninth of 54,000 equal 6,000 ($\frac{1}{9} \times 54,000 = 6,000$).

$(54,000 - 6,000 = 48,000)$ fans are attending this week

9) Answer: C

the population is increased by 25% and 20%. 25% increase changes the population to 125% of original population. For the second increase, multiply the result by 120%.

$(1.25) \times (1.20) = 1.50 = 150\%$

50 percent of the population is increased after two years.

10) Answer: B

Solve for $x \Rightarrow 6 \leq 2x + 4 < 20 \Rightarrow$ (add -4 all sides) $6 - 4 \leq 2x + 4 - 4 < 20 - 4$

$\Rightarrow 2 \leq 2x < 16 \Rightarrow$ (divide all sides by 2) $1 \leq x < 8$

x is between 1 and 8 Choice B represent this inequality.

11) Answer: B

To get a sum of 5 for two dice, we can get 4 different options:

$(1, 4), (4, 1), (2, 3), (3, 2)$

To get a sum of 7 for two dice, we can get 6 different options:

$(1, 6), (6, 1), (2, 5), (5, 2), (3, 4), (4, 3)$

To get a sum of 10 for two dice, we can get 3 different options: $(4, 6), (6, 4), (5, 5)$

Therefore, there are 13 options to get the sum of 5 or 7 or 10.

Since, we have $6 \times 6 = 36$ total options, the probability of getting a sum of 5 or 7 or 10 is 13 out of 36 or $\frac{13}{36}$.

WWW.MathNotion.Com

GED Subject Test – Mathematics

12) Answer: 12

Use formula of rectangle prism volume.

V = (length) (width) (height) ⇒ 6,144 = (32) (16) (height)

⇒ height = 6,144 ÷ 512 = 12

13) Answer: A

average (mean) = $\frac{\text{sum of terms}}{\text{number of terms}}$ ⇒ 64 = $\frac{\text{sum of terms}}{35}$ ⇒ sum = 64 × 35 = 2,240

The difference of 82 and 47 is 35. Therefore, 35 should be subtracted from the sum.

2,240 − 35 = 2,205

mean = $\frac{\text{sum of terms}}{\text{number of terms}}$ ⇒ mean = $\frac{2,405}{35}$ = 63

14) Answer: C

Change the numbers to decimal and then compare.

$\frac{1}{16}$ = 0.0625;   0.7;   25% = 0.25; $\frac{3}{4}$ = 0.75

Therefore $\frac{1}{16}$ < 25% < 0.7 < $\frac{3}{4}$.

15) Answer: B

Volume of a box = length × width × height = 10 × 7 × 5 = 350

16) Answer: B

average = $\frac{\text{sum of terms}}{\text{number of terms}}$ ⇒ 28 = $\frac{\text{sum of 9 numbers}}{9}$ ⇒ sum of 9 numbers = 28 × 9 = 252

35 = $\frac{\text{sum of 6 numbers}}{6}$ ⇒ sum of 6 numbers = 6 × 35 = 210

sum of 9 numbers − sum of 6 numbers = sum of 3 numbers

252 − 210 = 42

average of 3 numbers = $\frac{42}{3}$ = 14.

17) Answer: B

Solving Systems of Equations by Elimination

Multiply the second equation by (2), then add it to the first equation.

$\begin{matrix} 2x + y = -2 \\ 2(-x - y = -2) \end{matrix}$ ⇒ $\begin{matrix} 2x + y = -2 \\ -2x - 2y = -4 \end{matrix}$ ⇒ −y = −6 ⇒ y = 6

GED Subject Test – Mathematics

18) Answer: C

The width of the rectangle is twice its length. Let x be the length. Then, $width = 2x$

Perimeter of the rectangle is 2 (width + length) = $2(2x + x) = 84 \Rightarrow 6x = 84 \Rightarrow x = 14$; Length of the rectangle is 14 meters.

19) Answer: A and E

(In the stadium the ratio of home fans to visiting fans in a crowd is 5:7. Therefore, total number of fans must be divisible by 12: $5 + 7 = 12$. Let's review the choices:

A. 42,000: $42,000 \div 12 = 3,500$

B. 36,770 : $36,770 \div 12 = 3,064.17$

C. 52,300: $52,300 \div 12 = 4,358.33$

D. 61,750: $61,750 \div 12 = 5,145.833$

E. 33,000: $33,000 \div 12 = 2,750$

Only choices A and E when divided by 12 result a whole number.

20) Answer: 78

$\sqrt{380.25} = 19.5$

Four times the side of the square is the perimeter: $4 \times 19.5 = 78$

21) Answer: C

$\text{Probability} = \frac{\text{number of desired outcomes}}{\text{number of total outcomes}} = \frac{20}{24+36+20+30} = \frac{20}{110} = \frac{2}{11}$

22) Answer: D

Let x be the side.

Use Pythagorean Theorem: $a^2 + b^2 = c^2$

$x^2 + x^2 = 14^2 \Rightarrow 2x^2 = 196 \Rightarrow x^2 = 98 \Rightarrow x = \sqrt{98}$

The area of the square is: $\sqrt{98} \times \sqrt{98} = 98$

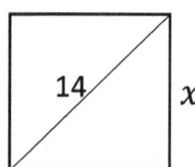

23) Answer: B and D

$6x - 2y = -10$. Plug in the values of x and y from choices provided. Then:

A. $(-3, -3)$: $6(-3) - 2(-3) = -18 + 6 = -12$, This is NOT true!

B. $(1, 8)$: $6(1) - 2(8) = 6 - 16 = -10$, This is true!

GED Subject Test – Mathematics

C. $(-1, -2)$: $6(-1) - 2(-2) = -6 + 4 = -2$, This is NOT true

D. $(0, 5)$: $6(0) - 2(5) = 0 - 10 = -10$, This is true!

24) Answer: A

To find the number of possible outfit combinations, multiply number of options for each factor: $3 \times 5 \times 10 = 150$

25) Answer: C

The ratio of boy to girls is 4:5. Therefore, there are 4 boys out of 9 students. To find the answer, first divide the total number of students by 9, then multiply the result by 4.

$72 \div 9 = 8 \Rightarrow 8 \times 4 = 32$

There are 32 boys and 40 (72 – 32) girls. So, 8 more boys should be enrolled to make the ratio 1:1

26) Answer: C

First, find the sum of three numbers.

Average = $\frac{\text{sum of terms}}{\text{number of terms}} \Rightarrow 64 = \frac{\text{sum of 3 numbers}}{3} \Rightarrow$ sum of 3 numbers = $3 \times 64 = 192$

The sum of 3 numbers is 192. If a fourth number that is greater than 78 is added to these numbers, then the sum of 4 numbers must be greater than 270 (192 + 78 = 270), then the average of these new numbers is:

average = $\frac{\text{sum of terms}}{\text{number of terms}} = \frac{270}{4} = 67.5$

Since the number is bigger than 78. Then, the average of four numbers must be greater than 67.5.

27) Answer: C

The perimeter of the trapezoid is 56 cm.

Therefore, the missing side (height) is = $56 - (16 + 14 + 12) = 14$

Area of a trapezoid: A = $\frac{1}{2}$ h (b$_1$ + b$_2$) = $\frac{1}{2}$ (14) (14 + 12) =182

28) Answer: C

The probability of choosing Hearts or Diamonds is $\frac{26}{52} = \frac{1}{2}$

GED Subject Test – Mathematics

29) Answer: 34

Let x be the width of the rectangle.

Use Pythagorean Theorem: $a^2 + b^2 = c^2$

$x^2 + 5^2 = 13^2 \Rightarrow x^2 + 25 = 169 \Rightarrow x^2 = 169 - 25 = 144$

$\Rightarrow x = 12$.

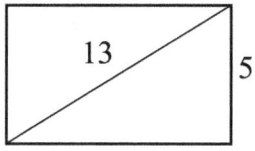

Perimeter of the rectangle = 2 (length + width) = 2 (12 + 5) = 34

30) Answer: A

Surface Area of a cylinder = 2πr (r + h),

The radius of the cylinder is 3 (6 ÷ 2) inches and its height is 12 inches. Therefore,

Surface Area of a cylinder = 2π (3) (3 + 12) = 90π

31) Answer: D

Simplify: $4x^3y^6(-3xy^2)^2 = 4x^3y^6(9x^2y^4) = 36x^5y^{10}$

32) Answer: B

3,500 out of 84,000 equals to $\frac{3,500}{84,000} = \frac{1}{24}$

33) Answer: C

Write the numbers in order: 10, 11, 16, 18, 19, 23, 29

Median is the number in the middle. So, the median is 18.

34) Answer: C

Use simple interest formula:

$I = prt$ (I = interest, p = principal, r = rate, t = time)

$I = (33,000)(0.035)(5) = 5,775$

35) Answer: 89

Mrs. Thomson needs an 90% average to pass for five exams. Therefore, the sum of 5 exams must be at lease $5 \times 90 = 450$

The sum of 4 exams is: $78 + 92 + 95 + 96 = 361$

The minimum score Mrs. Thomson can earn on her fourth and final test to pass is: $450 - 361 = 89$

GED Subject Test – Mathematics

36) Answer: D

The distance between Daniel and Noa is 35 miles. Daniel running at 4 miles per hour and Noa is running at the speed of 9 miles per hour. Therefore, every hour the distance is 5 miles less. $35 \div 5 = 7$.

37) Answer: B

Plug in 77 for F and then solve for C.

$C = \frac{5}{9}(F - 32) \Rightarrow C = \frac{5}{9}(77 - 32) \Rightarrow C = \frac{5}{9}(45) = 25$

38) Answer: A

Let x be the number of new shoes the team can purchase. Therefore, the team can purchase $92x$. The team had $52,000 and spent $20,500.

Now, write the inequality: $92x + 20{,}500 \leq 52{,}000$

39) Answer: A

Let x be the number. Write the equation and solve for x.

$60\%\ of\ x = 45 \Rightarrow 0.60x = 45 \Rightarrow x = 45 \div 0.60 = 75$

40) Answer: B

Let x be all expenses, then $\frac{20}{100}x = \$840 \rightarrow x = \frac{100 \times \$840}{20} = \$4{,}200$

He spent for his rent: $\frac{30}{100} \times \$4{,}200 = \$1{,}260$

41) Answer: D

The failing rate is 66 out of 165, $\frac{66}{165}$

Change the fraction to percent: $\frac{66}{165} \times 100\% = 40\%$

40 percent of students failed.

Therefore, 60 percent of students passed the exam.

42) Answer: 50

Use the information provided in the question to draw the shape. $a^2 + b^2 = c^2$

Use Pythagorean Theorem: $30^2 + 40^2 = c^2 \Rightarrow 900 + 1{,}600 = c^2 \Rightarrow 2{,}500 = c^2 \Rightarrow c = 50$.

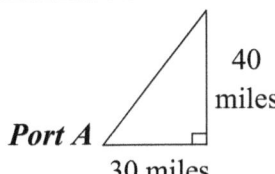

GED Subject Test – Mathematics

43) Answer: D

$\frac{75}{192}$, simplify by 3, then the number is the square root of $\frac{25}{64}$

$\sqrt{\frac{25}{64}} = \frac{5}{8}$; The cube of the number is: $(\frac{5}{8})^3 = \frac{125}{512}$

44) Answer: C

Isolate and solve for x.

$\frac{1}{5}x + \frac{4}{15} = \frac{1}{2} \Rightarrow \frac{1}{5}x = \frac{1}{2} - \frac{4}{15} = \frac{7}{30} \Rightarrow \frac{1}{5}x = \frac{7}{30}$

Multiply both sides by the reciprocal of the coefficient of x.

$\frac{1}{5}x = \frac{7}{30} \Rightarrow x = \frac{7}{6}$

45) Answer: A

First, find the number.

Let x be the number. Write the equation and solve for x.

125 % of a number is 85, then: $1.25 \times x = 85 \Rightarrow x = 85 \div 1.25 = 68$

45% of 68 is: $0.45 \times 68 = 30.6$

46) Answer: $-\frac{5}{8}$

Solve for y. $-5x - 8y = 20 \Rightarrow -8y = 20 + 5x \Rightarrow y = -\frac{5}{8}x - \frac{5}{2}$

The slope of the line is $-\frac{5}{8}$

GED Subject Test – Mathematics

Practice Test 2
GED Mathematical Reasoning

1) Answer: B

Plug in each pair of numbers in the equation: $x - \frac{7}{5}y = -3$

A. $(-5, -10)$: $(-5) - \frac{7}{5}(-10) = 9$

B. $(4, 5)$: $(4) - \frac{7}{5}(5) = -3$

C. $(-2, 5)$: $(-2) - \frac{7}{5}(5) = -9$

D. $(4, -5)$: $(4) - \frac{7}{5}(-5) = 11$.

2) Answer: A

To find the discount, multiply the number by (100% – rate of discount).

Therefore, for the first discount we get: (530) (100% – 22%) = (530) (0.78)

For the next 13% discount: (530) (0.78) (0.87).

3) Answer: D

$-2x + 5 = 9 \rightarrow -2x = 9 - 5 = 4 \rightarrow x = \frac{4}{-2} = -2$

Then; $-2x - \frac{3}{4} = -2(-2) - \frac{3}{4} = 4 - 0.75 = 3.25$.

4) Answer: −20

Use PEMDAS (order of operation):

$11 \times (-2) + 8 + 2(-15 - 7 \times 3) \div 12 = -22 + 8 + 2(-15 - 21) \div 12 = -14 + 2(-36) \div 12 = -14 - 72 \div 12 = -14 - 6 = -20$.

5) Answer: B

First draw an isosceles triangle. Remember that two sides of the triangle are equal. Let put a for the legs. Then: $a = 6 \Rightarrow$ area of the triangle is $= \frac{1}{2}(6 \times 6) = \frac{36}{2} = 18\ cm^2$.

Isosceles right triangle

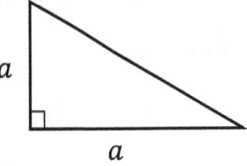

GED Subject Test – Mathematics

6) Answer: C

The average speed of Ryan is: $450 \div 9 = 50$ km

The average speed of Riley is: $490 \div 7 = 70$ km

Write the ratio and simplify.

$50 : 70 \Rightarrow 5 : 7$.

7) Answer: −1

Solving Systems of Equations by Elimination

$\begin{array}{l} 4x + 3y = 5 \\ -x - 2y = -5 \end{array}$ Multiply the first equation by 2, and second equation by 3, then add two equations.

$\begin{array}{l} 2(4x + 3y = 5) \\ 3(-x - 2y = -5) \end{array} \Rightarrow \begin{array}{l} 8x + 6y = 10 \\ -3x - 6y = -15 \end{array} \Rightarrow 5x = -5 \Rightarrow x = -1$.

8) Answer: C

The sum of supplement angles is 180. Let x be that angle. Therefore, $x + 9x = 180$.

$10x = 180$, divide both sides by 10: $x = 18$.

9) Answer: A

Use percent formula: Part = $\frac{percent \times whole}{100}$

$369 = \frac{percent \times 450}{100} \Rightarrow$ (cross multiply): $369 = percent \times 450 \Rightarrow percent = \frac{36,900}{450} = 82$

: $100\% - 82\% = 18\%$.

10) Answer: A

Let x be the number. Write the equation and solve for x.

$\frac{(90-x)}{x} = 5$ (cross multiply)

$(90 - x) = 5x$, then add x both sides. $90 = 6x$, now divide both sides by 6. $\Rightarrow x = 15$.

11) Answer: A

Use Pythagorean Theorem: $a^2 + b^2 = c^2$

$5^2 + 12^2 = C^2 \Rightarrow 25 + 144 = C^2 \Rightarrow 169 = c^2 \Rightarrow c = 13$

WWW.MathNotion.Com

GED Subject Test – Mathematics

12) Answer: D

$18 \times 10 = \$180$

Petrol use: $3 \div 2 = 1.5 \ per \ hour, 10 \times 1.5 = 15$ liters

Petrol cost: $15 \times \$3.25 = \$48.75$

Money earned: $\$180 - \$48.75 = \$131.25$

13) Answer: C

If the length of the box is 80, then the width of the box is one fourth of it, 20, and the height of the box is 2 (one tenth of the width). The volume of the box is: $V = lwh = (80)(20)(2) = 3,200$

14) Answer: D

Let x be the original price.

If the price of the sofa is decreased by 10% to $1,350, then: $90\% \ of \ x = 1,350 \Rightarrow 0.9x = 1,350 \Rightarrow x = 1,350 \div 0.90 = 1,500$

15) Answer: B

The percent of girls playing tennis is: $60\% \times 40\% = 0.60 \times 0.4 = 0.24 = 24\%$

16) Answer: 48

average $= \frac{sum \ of \ terms}{number \ of \ terms} \Rightarrow 44 = \frac{(42+46+40+x)}{4} \Rightarrow 176 = 128 + x \Rightarrow x = 48$

17) Answer: A

Use this formula: Percent of Change $= \frac{New \ Value - Old \ Value}{Old \ Value} \times 100\%$

$\frac{25,200-36,000}{36,000} \times 100\% = -30\%$ and

$\frac{25,200-17,640}{25,200} \times 100\% = -30\%$

18) Answer: C

Let x be the smallest number. Then, these are the numbers:

$x, x+1, x+2, x+3, x+4, x+5, x+6$

average $= \frac{sum \ of \ terms}{number \ of \ terms} \Rightarrow 57 = \frac{x+(x+1)+(x+2)+(x+3)+(x+4)+(x+5)+(x+6)}{7}$

$\Rightarrow 57 = \frac{7x+21}{7} \Rightarrow 57 = x + 3 \Rightarrow x = 54$

GED Subject Test – Mathematics

19) Answer: A and E

(If you selected 3 choices and 2 of them are correct, then you get one point. If you answered 2 or 3 choices and one of them is correct, you receive one point. If you selected more than 3 choices, you won't get any point for this question.)

Area of the circle is less than 81π. Use the formula of areas of circles.

$$Area = \pi r^2 \Rightarrow \pi r^2 < 64\pi \Rightarrow r^2 < 64 \Rightarrow r < 8$$

Radius of the circle is less than 8. Let's put 7 for the radius. Now, use the circumference formula:

$$Circumference = 2\pi r = 2\pi (8) = 16\pi$$

Since the radius of the circle is less than 8. Then, the circumference of the circle must be less than 16π. Online choices A and E are less than 16π

20) Answer: A

Use simple interest formula: $I = prt$

(I = interest, p = principal, r = rate, t = time)

$I = (12,800)(0.0225)(6) = 1,728$

21) Answer: B

Add the first 4 numbers. $56 + 42 + 68 + 24 = 190$

To find the distance traveled in the next 4 hours, multiply the average by number of hours.

$$Distance = Average \times Rate = 37 \times 6 = 222$$

Add both numbers. $190 + 222 = 412$

22) Answer: 64.

The ratio of boy to girls is 2: 7. Therefore, there are 2 boys out of 9 students. To find the answer, first divide the total number of students by 9, then multiply the result by 3.

$288 \div 9 = 32 \Rightarrow 32 \times 2 = 64$

23) Answer: C

Use distance formula: $Distance = Rate \times time$

$\Rightarrow 120 = 16 \times T$, divide both sides by 16. $\Rightarrow T = 7.5$ hours.

GED Subject Test – Mathematics

Change hours to minutes for the decimal part. $0.5\ hours = 0.5 \times 60 = 30\ minutes$.

24) The answer is 444.

The perimeter of the trapezoid is 86.

Therefore, the missing side (height) is $= 86 - (17 + 20 + 25) = 24$

Area of a trapezoid: $A = \frac{1}{2} h (b1 + b2) = \frac{1}{2} (24)(17 + 20) = 444$

25) Answer: C and E

The equation of a line is in the form of $y = mx + b$, where m is the slope of the line and b is the $y-intercept$ of the line.

Two points $(1, -3)$ and $(4, 6)$ are on the line A. Therefore, the slope of the line A is:

$slope\ of\ line\ A = \frac{y_2 - y_1}{x_2 - x_1} = \frac{6-(-3)}{4-1} = \frac{9}{3} = 3$

The slope of line A is 3. Thus, the formula of the line A is:

$y = mx + b = 3x + b$, choose a point and plug in the values of x and y in the equation to solve for b. Let's choose point $(1, -3)$. Then:

$$y = 3x + b \to -3 = 3 + b \to b = -3 - 3 = -6$$

The equation of line A is: $y = 3x - 6$

Now, let's review the choices provided:

A. $(5, 4)$ $y = 3x - 6 \to 4 = 15 - 6 = 9$, This is not true.

B. $(-1, 7)$ $y = 3x - 6 \to 7 = -3 - 6 = -9$, This is not true!

C. $(2, 0)$ $y = 3x - 6 \to 0 = 6 - 6 = 0$, This is true.

D. $(-3, 4)$ $y = 3x - 6 \to 4 = -9 - 6 = -15$, This is not true!

E. $(-2, -12)$ $y = 3x - 6 \to -12 = -6 - 6 = -12$, This is true.

26) Answer: B

Use percent formula: $Part = \frac{percent \times whole}{100}$

$40 = \frac{percent \times 32}{100} \Rightarrow \frac{40}{1} = \frac{percent \times 32}{100}$, cross multiply.

$4{,}000 = percent \times 32$, divide both sides by 32.

$125 = percent$

GED Subject Test – Mathematics

27) Answer: B

To find the discount, multiply the number by (100% − rate of discount).

Therefore, for the first discount we get: $(100\% - 20\%)(E) = (0.80)E$

For increase of 5 %:

$(0.80)E \times (100\% + 5\%) = (0.80)(1.05) = 0.84E$.

28) Answer: C

The weight of 25 meters of this rope is: $25 \times 340g = 8,500g$

1 kg = 1,000 g, therefore, $8,500\ g \div 1,000 = 8.500 kg$

29) Answer: A and C

(If you selected 3 choices and 2 of them are correct, then you get one point. If you answered 2 or 3 choices and one of them is correct, you receive one point. If you selected more than 3 choices, you won't get any point for this question.)

Some of prime numbers are: 2, 3, 5, 7, 11, 13, 17, 19

Find the product of two consecutive prime numbers:

$7 \times 3 = 21$ (bingo!)

$11 \times 5 = 55$ (not in the options)

$13 \times 11 = 143$ (not in the options)

$17 \times 13 = 221$ (yes!)

Choices A and C are correct.

30) Answer: C

Use the formula for Percent of Change: $\frac{New\ Value - Old\ Value}{Old\ Value} \times 100\ \%$

$\frac{69-120}{120} \times 100\ \% = -42.5\ \%$ (negative sign here means that the new price is less than old price).

31) Answer: B

If the score of Harper was 189, therefore the score of Emma is 63. Since, the score of Zoe was one fifth of Emma, therefore, the score of Zoe is 12.6.

32) Answer: B

Let x be the number. Write the equation and solve for x.

GED Subject Test – Mathematics

$\frac{3}{7} \times 63 = \frac{9}{11} \cdot x \Rightarrow \frac{3 \times 63}{7} = \frac{9x}{11}$, use cross multiplication to solve for x, $33 \times 63 = 9x \times 7 \Rightarrow 2{,}079 = 63x \Rightarrow x = 33$

33) Answer: A

If 15 balls are removed from the bag at random, there will be five ball in the bag. The probability of choosing a blue ball is 5 out of 20. Therefore, the probability of not choosing a white ball is 15 out of 20 and the probability of having not a white ball after removing 15 balls is the same ($\frac{5}{20} = \frac{1}{4}$).

34) Answer: 729

$9^3 = 9 \times 9 \times 9 = 729$

35) Answer: A

Write the **numbers** in order: 7, 11, 14, 19, 20, 20, 24, 26, 27

Since we have 9 numbers (9 is odd), then the median is the number in the middle, which is 20.

36) Answer: C

The area of the floor is: $5\ cm \times 56\ cm = 280\ cm^2$

The number of tiles needed $= 280 \div 8 = 35$

37) Answer: B

6% of the volume of the solution is alcohol. Let x be the volume of the solution. Then:

6% of $x = 36.6$ ml; $0.06x = 36.6 \Rightarrow \frac{6x}{100} = \frac{366}{10}$ cross multiply

$60x = 36{,}600 \Rightarrow$ (devide by 60) $x = 610$

38) Answer: 602.88

Surface Area of a cylinder = $2\pi r\ (r + h)$,

The radius of the cylinder is 6 inches and its height are 10 inches. π is 3.14. Then:

Surface Area of a cylinder = 2 (3.14) (6) (6 + 10) = 602.88

39) Answer: A

Average = $\frac{sum\ of\ terms}{number\ of\ terms}$

The sum of the high of all constructions is: $24 \times 135 = 3{,}240$ m

WWW.MathNotion.Com

GED Subject Test – Mathematics

The sum of the high of all towers is: $16 \times 165 = 2,640$ m

The sum of the high of all building is: $3,240 + 2,640 = 5,580$

Average = $\frac{5,580}{40} = 147$

40) Answer: A

Let x be the original price.

If the price of a laptop is decreased by 40% to $615, then:

$60\% \text{ of } x = 615 \Rightarrow 0.60x = 615 \Rightarrow x = 615 \div 0.60 = 1,250$

41) Answer: B

Let x be the number of years. Therefore, $3,420 per year equals $3,420x$.

starting from $48,000 annual salary means you should add that amount to $3,420x$.

Income more than that is: $I > 3,420x + 48,000$

42) Answer: C

Use the information provided in the question to draw the shape.

Use Pythagorean Theorem: $a^2 + b^2 = c^2$

$9^2 + 12^2 = c^2 \Rightarrow 81 + 144 = c^2 \Rightarrow 225 = c^2$

$\Rightarrow c = 15.$

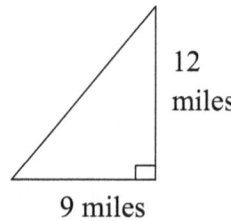

43) Answer: A

Write the equation and solve for M:

0.96 F $= 0.16$ M, divide both sides by 0.16, then: $\frac{96}{16}$ F $=$ M, therefore:

M $= 6$ F, and M is 6 times of F or it's 600% of F.

44) Answer: B

For each option, choose a point in the solution part and check it on both inequalities.

$$y \leq x + 3$$
$$x + 3y \geq 6$$

A. Point $(-1, 3)$ is in the solution section. Let's check the point in both inequalities.

$3 \leq -1 + 3,$ That's not true

$-1 + 3(3) \geq 6 \Rightarrow 8 \geq 6,$ That's true

WWW.MathNotion.Com

GED Subject Test – Mathematics

B. Let's choose this point (2, 2); $2 \leq 2 + 3$, That's true

    $2 + 3(2) \geq 6$, That's true!

C. Let's choose this point (0, 0); $0 + 3(0) \geq 6$, That's not true!

D. Let's choose this point (–4, 1); $1 \leq -4 + 3$, That's not true!

45) Answer: A

The question is this: 4.8 is what percent of 1.2?

Use percent formula: part $= \frac{percent}{100} \times whole$

$4.8 = \frac{percent}{100} \times 1.2 \Rightarrow 4.8 = \frac{percent \times 1.2}{100} \Rightarrow 480 = percent \times 1.2 \Rightarrow percent = \frac{480}{1.2} = 400$.

46) Answer: 216

To find the number of possible outfit combinations, multiply number of options for each factor: $6 \times 4 \times 9 = 216$

"End"